HIGH-TECH CAREER STRATEGIES FOR WOMEN

ALSO BY JOAN RACHEL GOLDBERG

You CAN Afford a Beautiful Wedding

HIGH-TECH CAREER STRATEGIES FOR WOMEN

Joan Rachel Goldberg

MACMILLAN PUBLISHING COMPANY

New York

Macmillan Publishing Company
866 Third Avenue, New York, N.Y. 10022
Collier Macmillan Canada, Inc.

Library of Congress Cataloging in Publication Data

Goldberg, Joan Rachel.
High-tech career strategies for women.

Bibliography: p.
Includes index.
1. High technology industries—Vocational guidance.
2. Vocational guidance for women. 3. Job hunting.
I. Title.
HC59.G615 1984b 621.381'7'02373 84-15426
ISBN 0-02-544460-3

10 9 8 7 6 5 4 3 2 1

Designed by Jack Meserole

Printed in the United States of America

Contents

Acknowledgments

I am very grateful to the many people who helped me research and write this book. In particular, I want to thank my agent, Barbara Bova. Without her efforts, this book would not exist. Similarly, the interest and ideas of Alexia Dorszynski helped make this book a reality.

My thanks, too, to the following people who shared their thoughts and offered their encouragement and support: Caryl Avery, Fiona Baxter, Mark and Temma Berg, Beth Brophy, Pat Carretta, Marcy Ersoff, Susanne Feld, Rose Ann and Neil Fraistat, Deborah Goldberg, Elaine and Jack Goldberg, Anne Guthrie, Arthur Karlin, Ada and Julie Leinwand, Marty Leinwand, Paul Leinwand, Fran Lerner, Beth and Bill Loizeaux, Janice Miller, Al Munzer, Lenore Salzbrunn, Andy Schoenholtz, Joel Swetow, and Vivian Wishingrad.

In addition, I want to thank everyone whom I interviewed for this book. There are many names scattered throughout the following pages: my gratitude to all of these high-tech women and men, psychologists and career counselors, industry analysts and venture capitalists, journalists and researchers for graciously sharing their stories and their advice.

Special thanks to Laura Greenberg for her inventive ideas and for being there. And, as always, I consider myself lucky to have received the support, encouragement, love—and editorial assistance—of my husband, Ted Leinwand. Without him, I never would have completed this book in such good spirits.

HIGH-TECH CAREER STRATEGIES FOR WOMEN

An Overview

What are high-tech industries? What exactly is the much talked about information revolution? How have computers changed the face of the labor market? How has technology caused dramatic career shifts for so many? Is the high-tech world wide open to women? And are there golden opportunities for women with technical training, or for *all* women with drive and ambition? These are just some of the questions this book will answer.

Whatever your present status—manager or secretary, technician or trainer, student or homemaker—you're probably curious about the potential a high-tech career offers. Whether you've programmed computers or merely read computer printouts, whether you understand the technology behind videotex (more on this later) or fear what you don't understand, high-tech careers offer you terrific potential. Computers and other forms of technology have changed the structure of the work world.

As computers have flooded the offices of the western world, the ranks of middle management have shrunk. (Computers facilitate information gathering and decisionmaking, two prime functions of middle managers.) Secretaries rarely just type reports today; they word process them. And in accepting computers or word processors and other new technology into their offices, secretaries have the opportunity to learn something new—and to capitalize on their knowledge. Many secretaries have used computer know-how as a stepping stone to training and managerial positions. Other secretaries have discovered that computers can make their jobs easier—or non-existent. Similarly, drafters may find their jobs in jeopardy as computer-aided design and manufacturing becomes more competitive.

Factory workers may welcome robots on the assembly line, or feel threatened by them. Those who work with numbers find their work simplified by computers—which makes computer skills a job requirement for accountants, financial planners, and other statistical workers. Keypunch operators are following in the steps of dinosaurs—toward extinction. And programmers, too, are finding that their jobs change dramatically as technology advances.

Artists may turn to computer graphics, writers to software designing. Teachers may switch to computer literacy training or to developing educational software. Salespeople may find greater rewards as well as greater challenges in selling high-tech products. And managers may find that technology simplifies many procedures, while adding new tasks to their job descriptions.

Technology triggers change, and many people—unfortunately—are frightened by change. But technology is here to stay. And what is high-tech today will probably become low-tech tomorrow. Just as the railroad changed transportation forever—and the telephone transformed communication—computers, robots, biotechnology, CAD/CAM (computer-aided design, computer-aided manufacturing), electronic mail, fiber optics, medical imaging, lasers, telecommunications, and artificial intelligence (AI) are changing our world today. High technology simply refers to the cutting edge of technological development. And since technology advances so rapidly, much of what is labelled high-tech today really shouldn't be. Even though computers have been around since the 1940s (in much less sophisticated form!), they are called high-tech because many of their eighties *applications* are new.

Change isn't new to the 1980s, and neither is adaptation to change. It's the speed of technological change that leaves many of us dizzied and confused. Still, that's no excuse for failing to pay attention to the career potential high-tech industries offer. Four of the U.S.'s five fastest-growing jobs in the next decade will be high-tech. And seven of the ten fastest-growing *areas* will involve computers or engineering. While it's also true that many new jobs won't be high-tech, the fact remains that high-tech jobs may well offer you the greatest opportunities for career growth, intellectual stimulation, financial rewards, and flexibility. High-tech careers aren't the only game in town, but they may well be the most worthy of your consideration. Are you just embarking on a career path? Or are you dissatisfied with your present status and interested in making a smart career move—one that will satisfy your present career needs and pay off in the long run?

High-tech fields offer both men and women valuable career opportunities. Yet too frequently women fear—or are told—that they lack sufficient technical skills or abilities to jump on the high-tech bandwagon. Chapter 2 explains why this idea is false. And it details research indicating just the contrary, that women may well have an *advantage* in high-tech careers.

If you're uncertain about what the high-tech world offers you in terms of career satisfactions and benefits, read this book carefully. You'll hear the stories of women who have found career contentment—and excitement—in high-tech positions. You'll discover how you can get training (if you need it), what and where the jobs are, how to select the best job for you and how to land it, how to move and advance in a high-tech career, and how to start and run a successful high-tech business.

First, let's start with the stuff dreams are made of: success stories. In the following pages you'll read about high-tech women who've made it. Many of them are entrepreneurs. Some, no doubt, are women you've heard of—like astronaut Sally Ride and U.S. Navy Commodore Grace Hopper. All are women you should know about. Why? Because they've done it, they've pioneered various high-tech career paths. And that makes them high-tech role models. Because they happen to be women, you may feel especially inspired by their examples.

These women may have faced obstacles like sex discrimination or lack of technical or business training. They may have had to combat personal barriers, like fear of success or even technophobia. These women aren't myths. They're real women who are (in the case of Ada Augusta, were) smart, ambitious, dedicated, and successful. And they haven't traded humor, compassion, or their right to a personal life for success. They work hard and their examples point out possible routes for you to follow.

Role models are important because they suggest possibilities. They should not limit your ambitions, but rather expand them. And if you want to feel proud of their accomplishments, I'm sure they wouldn't mind!

ADA AUGUSTA

The first programmer was a gifted mathematician and translator who lived long before the Univac and the Eniac were invented. Ada Augusta, Lady Lovelace, lived in the 1800s and is frequently referred to as the first programmer because of her collaboration with inventor Charles Babbage. Babbage, far ahead of his time, was responsible for the first "computer." And Lady Lovelace devised a system for betting on horses using Babbage's computing machine. (It failed, but then so do many software packages today before they are sufficiently debugged, or freed of errors.)

Lady Lovelace was the daughter of Anne Milbanke, a mathematician, and Lord Byron, the poet. (In fact, Byron once called Milbanke—his first wife—the "Princess of Parallelograms.") Ada Augusta died in 1852 at the age of 37.

Her story illustrates women's early involvement in computer history and indicates what has since become an old, tried and true, high-tech career route: the study of mathematics and languages. And working partnerships are still a common means of operating in the high-tech world. Just as computers require programs, computer designers are complemented by software writers. There is room in the high-tech world for many kinds of talents and many different sorts of creativity.

GRACE HOPPER

Skipping forward a century, we come to the story of Grace Hopper, who is now a commodore in the U.S. Navy. At seventy-seven, she is the Navy's oldest officer on active duty. She has been called "a walking textbook" and "the grand old lady of software." Both labels are accurate. Commodore Hopper is an amazing woman.

"Don't let anyone say, 'but we've always done it that way,' " she once admonished a young Navy officer. The statement could well be her personal motto. After earning her B.A. from Vassar in 1928, and her Ph.D. in mathematics from Yale in 1934, Hopper taught mathematics at Vassar and Barnard until 1944. There was a war on, notes Hopper, and she contributed to the war effort as a mathematical officer in the U.S. Navy from 1944 to 1946. It was during this time that Hopper, at the age of thirty-eight, encountered her first computer. "Don't let your age bother you," she advises others. "You have to be willing to work at learning how to use computers. The computer is not a thing in itself," says Hopper. "It's a tool."

Hopper programmed the Mark I, which was the first, large-scale American computer. She also developed the first practical compiler, and created COBOL, still one of the most common business computer languages. The term "bug," which indicates a computer glitch or problem in a program, was coined by Hopper and colleagues after they discovered that a moth in the circuit was the cause of the Mark I's malfunctioning.

Hopper's illustrious career includes research and teaching positions at Harvard, the University of Pennsylvania, and George Washington University, as well as technical and administrative positions at the Univac division of Sperry (where 50 percent of Univac's employees during World War II were women, Hopper remembers). She was called to active duty at the Navy in 1967, a year after she had retired from the Navy Reserve. The Navy needed someone to standardize its COBOL program and they went directly to the source, its inventor. Hopper returned to her beloved Navy, and has remained there ever since. In 1973, when she was too old for a regular promotion, a special act of Congress promoted Hopper to the rank of captain. On December 15, 1983, Hopper was promoted to commodore by President Reagan.

As for women programmers, Hopper prefers them to men. "They have, innately, a tendency to finish things," she believes. "To document a program completely and wrap it up in a nice package. This comes from our backgrounds."

Yet background alone cannot guarantee high-tech career success. "You have to keep up with what's going on," Hopper warns. "The day you stop learning is the day you start to die." Her advice to women interested in high-tech careers is to get the best training possible. And "set out to be a character," she adds, laughing. Hopper is aware that she's a role model. And now that she is a commodore, she knows she has to be a little more careful about what she says. "I must learn to be more dignified," she says with a twinkle in her eyes. Yet what she claims to be more significant in high-tech careers is having a sense of humor. Working with computers is "lots of fun."

Today Hopper travels around the country and abroad, lecturing to people of all ages about the excitement in store for those who enter the high-tech world. And after a lifetime of awards and a long list of publications, Commodore Grace Hopper surely deserves the title "grand old lady of software."

RUTH M. DAVIS

Ruth M. Davis is some twenty odd years Hopper's junior, but her accomplishments in the computer industry are also quite remarkable. Like Hopper, Davis began her career as a mathematician. She earned her bachelor's, master's, and doctoral degrees in mathematics, all *summa cum laude*. Yet Davis's professional life has been spent largely in government work, not in academia. And her skills at organization have been as highly lauded as her administrative talent and technical skills. She is perhaps best known for having organized the development of ultrafast integrated circuits at the U.S. Defense Department. She has also directed the department's research into particle-beam weapons. Davis has worked with Admiral Hyman Rickover (now retired) on developing the first computer programs used to design nuclear power plans. Other Davis accomplishments include successful promotion of a worldwide data encryption standard, furthering of the American position in the field of robotics, and implementation of a system in remote Alaska whereby the sick can be linked with outside doctors, via a satellite hookup.

Davis was director of the National Bureau of Standards' Institute of Computer Sciences and Technology, before leaving to join the Defense Department. The positions she has held have been called by *Datamation* "arguably the most sensitive ever held by a mathematician, a computer expert, or a woman." Like Hopper, Davis has been the recipient of a rich assortment of awards and honors.

After her sterling career in the U.S. government, Davis went on to found her own company, Pymatunning Group. She also serves on the boards of overseers for mathematics at Harvard and Dartmouth, and in a senior advisory position at the University of Pittsburgh. Davis also finds time for occasional stints as a lecturer on college campuses.

Ruth M. Davis's career illustrates the kind of drive and dedication needed to advance to such high levels in the high-tech—and government—worlds. And the solid grounding of an excellent education helps, as do top organizational and management skills.

Once an individual rises to management level in the high-tech world, technical expertise counts for less than managerial talent. (More on this in Chapter 7.) Davis's story teaches a truth about high-level career paths in the high-tech world. To rise far, one must eventually take on organizational and managerial roles. Not every high-tech individual has the potential for managerial success. However, high-tech industries do offer two different career paths for the upwardly mobile: one technical, and one managerial. Which route interests you?

EVELYN BEREZIN

Evelyn Berezin started her high-tech career in the private sector, and has remained there ever since. She's one of the pioneers in office automation, and her career illustrates the very meaning of the term risk-taker. In 1951 she graduated from New York University with a degree in physics. She stayed at her first company for six years, working as a computer designer in Brooklyn, New York. Her next job was at Digitronics Corporation, where Berezin worked on research and development of early automated word processing systems.

In 1969, Berezin left to start her first company, Redactron, with two colleagues. The impetus? "I had been working a long time," remembers Berezin, "and I noticed I wasn't going anywhere and I sure wasn't getting any richer. I had nothing to lose and everything to gain."

Was Berezin's lack of satisfactory career growth in the 1950s and 1960s due to sex discrimination or simply the nature of being an employee, not an entrepreneur? Berezin doesn't talk about discrimination: "I had worked in a man's world as a computer-design engineer for twenty-odd years. When I left Digitronics, I knew exactly where to go, what to do, and whom to contact. Twenty years gives a lot of opportunities for networking, and the industry was a very promising one. At that time it was wide open for anyone, man or woman, to do anything."

As with any new venture, Redactron could have been a success or failure. Thanks to Berezin's technical experience and business know-how, the company was a success. It went public in 1971, and grew from nine employees to 900 by 1976. In the late seventies Redactron was acquired by Burroughs and became a leading word-processing manufacturer. Berezin joined Burroughs as president of Burroughs Office Products Group in 1978. By 1979 she had left to start another new company, consulting on automation. And Berezin's eighties move was to form yet another company, Greenhouse Management Corporation. Greenhouse is a venture capital firm that invests in an area Berezin knows intimately—high-technology businesses. When Berezin started Greenhouse, she was also quite familiar with the workings of venture capitalists because of her own experience raising capital for Redactron. She understood business plans, budgets, and how to target goals. Greenhouse, like Redactron, was founded with business know-how and industry-specific skills.

Berezin advises women considering starting high-tech businesses to think big and to take risks: "Know deep in your heart that there is no security *anywhere*. It's an illusion, and one that women cling to. The only security is the strength of a new industry like office automation, plus your knowledge of your own abilities. If you feel you have something to market, go for it," she advises. And that's perhaps the best counsel one can ask for from a successful high-tech entrepreneur, which is what Evelyn Berezin is—three times over!

ELAINE BOND

Elaine Bond is senior vice-president and director of corporate information systems at Chase Manhattan Bank in New York City. After 23½ years at IBM, Bond took the difficult step of leaving IBM to move to Chase in 1981. She was one of four women that Catalyst, a non-profit women's organization, honored in 1983 for reaching the senior executive level in American corporations.

Bond began her career after earning a B.S. in math from Tufts University. Her first job was as a researcher/programmer-trainee at IBM. Successive moves made her a technical manager, and then data processing group director. In her present position at Chase, Bond manages 350 people. During her career, Bond has consciously shifted back and forth from line to staff positions (line positions contribute directly to a company's bottom line and are generally considered more essential, while staff positions tend to be in areas like personnel, public relations, advertising, and legal affairs) to broaden her experience. She counsels high-tech women to seek "depth first, but then breadth." Bond took on a position at IBM in human resources to further qualify her for future high-level corporate positions.

"The opportunities for leadership are very strong," she says. And for women in high-tech industries—especially in areas like sales, finance, and legal affairs—"you get to demonstrate proficiency very quickly." If you're good at what you do, "people tend to come to you," believes Bond.

As Davis's career illustrated, a woman needs skills beyond technical expertise to move to the senior level of a corporation or of government. Bond observes that the greatest adjustment for those with technical skills will be the realization that interpersonal skills and leadership ability become more important than technical expertise as an individual rises in the high-tech world.

Therefore Bond advises, "Force yourself to develop proficiency in other things around you."

Elaine Bond's career is unusual in the high-tech world because of the length of time she stayed with one corporation. More women than men have exhibited this kind of corporate loyalty (see Chapter 2 on how this can be an advantage), but this is not always beneficial to women's careers. Loyalty is not always rewarded sufficiently. Often an individual must move outside the company where she has started her career to advance to higher levels. Too frequently, individuals become "typecast" or stereotyped in the roles they hold. Company presidents or directors may become so accustomed to seeing an employee handle one position competently that they may be either too happy with the status quo or too unwilling to risk change. When an individual reaches the end of her career advancement at a particular company, she must decide whether she is happy enough where she is or ready to look elsewhere. In Elaine Bond's case, Chase came courting her. Leaving IBM was difficult for her because she was not unhappy. Yet the offer Chase made her was too good to refuse.

Bond, like all the role models profiled in this chapter, knows how to recognize a worthwhile challenge—and meet it. Banking is a conservative, traditionally male-dominated industry. Bond's career illustrates that women can succeed in such an industry and that a first (or even second or third) position in another industry does not preclude switching gears later.

High-tech positions cross a full range of industries. Technical and administrative positions in data processing and information systems are found in practically all industries. While some career experience is industry-specific—for example, programming for business applications will probably not be viewed as acceptable background for designing software for scientific use—much data processing and MIS (management of information systems) experience is basic to most industries. Working environments will change; so will the rules, written or unwritten, of employee interaction. For example, workers at software companies tend to be more informal than those at older, more established hardware companies, or in conservative industries like banking.

Bond's story indicates that personal ambition, and a well-managed career spent acquiring necessary experience and proving one's superior leadership ability, are necessary prerequisites for attaining top, high-tech, corporate positions. Bond is one of the select group of top-level female executives in corporate America. It's not a closed group, but it may be tougher to crack than the ranks of senior management of younger, less established, more liberal high-tech industries, like software. The choice is up to you. Elaine Bond is only one of the admirable role models you may choose to inspire your high-tech career direction.

JULIA WALSH

Julia Walsh's highly successful career is also highly unusual. She began, twenty-six years ago at the age of thirty-four, by investing all of her money in four stocks: Texas Instruments, Hewlett-Packard, Aerojet General, and Reynolds Metals. The risk was tremendous. Her husband had died, leaving Walsh with four young children and no additional money. The $20,000 she invested increased in worth fourfold within eighteen months. Walsh bought a house in Washington, D.C., where she lives today, and embarked on a new career. Today she runs her own investment firm, Julia M. Walsh & Sons, Inc., perhaps the only woman on the East Coast in such a position.

Walsh is on the boards of Pitney Bowes, Esmark Inc., and other companies. She has appeared regularly on the "Wall Street Week" television program. And her civic involvements include being a member of a Washington, D.C. real estate board and task force on housing. Walsh worked for the Democratic National Finance Committee and was considered by President Carter for the cabinet position of secretary of commerce. When Walsh was not appointed to the position, she decided to start her own firm —a place where her four children could learn her business and work with her. At sixty, Walsh heads an investment firm with thirty-five employees, about 4,000 accounts, and a seat on the New York Stock Exchange.

Walsh majored in international business at Kent State. Despite a lack of encouragement from her parents, Walsh pursued a career course not considered appropriate for females at the time. She rejected teaching and the arts, and after graduation joined the Foreign Service as a personnel officer. After three years she married her first husband and was kicked out of the Foreign Service as a result. Ten years and various types of employment later, Walsh was pregnant and decided the time was ripe to start business school at George Washington University. She loved studying financial statements and annual reports. And her talent for finances so impressed her teacher, George M. Ferris, Jr., that he offered her a position as a stockbroker in his securities firm, Ferris & Company, Inc.

In 1955, when Walsh was hired as a registered representative, there were few women stockbrokers. She was the only woman broker in the office and felt keenly the attitude that she wasn't serious. Her response: "I didn't do things that weren't ladylike. I accommodated the attitude that I wasn't very serious. I tried not to step on anyone's toes." She made safe investments for clients who wanted them, when she really wanted to be more adventurous. It wasn't until the late 1960s that Walsh began to follow her own head and invest in riskier high-tech stocks.

In the 1950s Walsh had to accommodate the prejudices of both clients and co-workers. Ferris sent her to an educational program for a week, where for the first few days all Walsh's colleagues—male—refused to sit next to her. When she told a colleague, "I don't think I can take this," he talked to the other men, and the next day the men sat next to her and talked to her. Unfortunately, this was not the last of the male chauvinism she experienced. At a Harvard thirteen-week advanced management program in 1962, many men resented the presence of a woman and made their resentment known. Walsh persevered.

At Ferris, where Walsh remained for twenty-two years, she moved up to vice-president. She taught at Catholic University and the Y; she spoke to Navy and Foreign Service groups in an aggressive attempt to promote herself and develop her business. In 1972, she became one of the few women to sit on the board of the American Stock Exchange. In 1977, when she started Julia M. Walsh & Sons, Inc. (her four oldest children are boys; her daughter is still in school), she had been earning more commissions than any of the 50 executives at Ferris.

Today Julia Walsh is worth nearly $3 million. She is successful, influential, and she's reached her present position without the assistance of the old-boy network. Walsh always knew she wanted to have a career that offered excitement and growth. And she decided to pass on that opportunity to her children, and to other employees.

The American image of stockbrokers has long been of stuffy, upper crust, Wall Street men who wear Brooks Brothers suits and bear degrees from Ivy League schools. Walsh doesn't fit that image. She doesn't play golf, and she doesn't belong to old-line clubs. She's earned her success because of her market savvy, hard work, and charisma. Yet as hard as she works, there's always time to enjoy her family (her second husband, of twenty-one years, is the vice-chairman of the board of National Permanent Federal Savings and Loan; together they have twelve children) and the things money can buy. "I am a terrible spendthrift," Walsh admits. "But, you know, you can't take it with you." Julia Walsh combines a drive for success with a keen sense of priorities, and for both of these reasons, she is an excellent role model for high-tech women.

LORE HARP

Lore Harp is well-known in high-tech circles for her entrepreneurial success. In 1976, at the age of thirty-two, she was bored as a housewife and decided to start a business with friend Carole Ely. Their product: a memory board (which stores information) developed by Harp's then husband, Bob Harp. Lore Harp knew little about technical matters, but she was determined to meet the challenge of starting a successful business. She and Ely attended the Southern California Computer Society Show to investigate the market for their product. The response was overwhelming.

"The hobbyists were very enthusiastic about computers, and very, very bright. They would peel off $800, $900, $1,000 in cash to buy a few memory boards. And I thought, 'My God, there must be a tremendous market.' " Within days after the show, Harp and Ely incorporated and focused their attention on marketing, advertising, packing, and shipping out their product—turning part of Harp's home into a combination office, factory, and warehouse. They brought in an accountant to set up their books, printed attractive flyers, designed a company logo, used quality paper, and took out ads in trade publications—initiating the company with a professional image. The boards were even packaged in attractive boxes printed with the company's logo in green and white.

"They used to call us the girls in green and white," Harp remembers. And while the term may well sound condescending, the company's growth was phenomenal. Harp and Ely worked long hours, calling dealers personally to make sales pitches. And they concentrated on providing excellent service—products shipped on time, good documentation with technical specifications and manuals. They even did trouble-shooting over the phone. (When customers reported problems or complaints that Harp and Ely were unable to answer, they took the questions to

Bob Harp.) Customers were impressed, dealers were impressed, and so were venture capitalists.

Vector Graphic Inc. was started with an investment of $12,000 (half from Harp, half from Ely). Six years later, in 1982, sales totalled $36 million. The history of the company has included tremendous growth, quick expansion, accepting venture capital, going public (selling stock in the company to the public to raise cash for the company), *and* difficult times. Through it all, Lore Harp, as founder, chief executive officer (CEO), and chairman of the board, has earned the respect of management consultants, venture capitalists, and industry analysts.

Harp's father was a businessman, but Harp herself had little business experience before founding Vector Graphic. She had a B.A. in anthropology, had attended law school for a year, had stayed home with her two children, and had worked outside the home only once—for six months in the international department of a bank. What Harp lacked in business experience, she made up in drive and leadership ability.

"You never say no to an opportunity," says Harp. While running Vector Graphic, Harp has earned an MBA from Pepperdine University in Los Angeles. Why? In business school, you study case histories and "learn other options." Harp knew the way her company made decisions; she learned there were other ways.

Manufacturing is "more of a grubby business" than other aspects of high-tech industries, believes Harp. It's also a very male-dominated business, more so than fields like data processing or computer publishing. Harp may have been more visible in the high-tech world because of her sex, but she says it hasn't been an advantage. She's aware that her position on the board of advisers of the professional society, Women in Information Processing (WIP; for more information, see Appendix I), on the board of directors for the American Electronics Association (also listed in Appendix I), and on the advisory board of the First Woman's Bank in Los Angeles add to her visibility. Yet visibility isn't a matter of concern for a company head. These positions are part of the professional responsibilities she shoulders.

Lore Harp was born in Germany. She left in 1966 (at the age of twenty) to visit friends in California. She decided to stay for a year, but changed her mind and has remained in this country ever since. (Harp's background as an immigrant is not an uncommon one for entrepreneurs. More on this in Chapter 8.)

Like many other gifted entrepreneurs, Harp seems a natural leader. She knows how to make fast decisions—a must in the rapidly changing world of advanced technology. And she's been able to ride with the changes that have brought many other high-tech companies to their knees. However, in April 1984, Harp left Vector Graphic to take over the presidency of Pacific Technology Venture Fund, Inc. in San Francisco.

The future of Vector Graphic and Pacific Technology cannot be predicted, but Lore Harp has high hopes. And she has proven her entrepreneurial mettle. For women who lack technical *and* business skills or experience, Harp's story illustrates that success is possible without them. Many entrepreneurs learn on the job. And while Harp may not consider herself a role model, many other people do—with good reason!

LENORE SALZBRUNN

Lenore Salzbrunn is another noteworthy high-tech entrepreneur. PEAR Systems Corporation, in Stamford, Connecticut, was the brainchild of Salzbrunn and partner Greg Wilson. Wilson wrote a program for portfolio management; Salzbrunn provided the business know-how for marketing the product, which was designed for Apple computers. The company, formed in June 1980, was sold in October 1982 to Remote Computing Corporation. As part of the deal, Salzbrunn joined the company, and continued to earn royalties on PEAR products. (Wilson became a consultant to Remote Computing.)

Today Salzbrunn works out of three offices—in Roslyn, New York, Manhattan, and her Connecticut home. Remote wanted to buy out PEAR Systems because Remote had been primarily a time-sharing business and wanted to expand into microcomputing. (Remote has since gone public, changing its name to Hale Systems, Inc.) Salzbrunn's present high-level executive position involves a great deal of travel for sales and promotion. She attends corporate meetings for planning and budgeting as well.

Salzbrunn has a two-year degree from a community college. Like many women of her generation, she stayed home to raise her children for ten years after her marriage. When she decided to return to the work force, she took jobs as an administrative assistant for an oil and gas investment firm and as a travel agent. Then Salzbrunn decided she wanted to run her own business. "I felt I had run enough businesses for other people, and I was interested in doing it for myself. So I decided very clearly I was going to do something. And probably making those conscious decisions is the best thing that can happen to anybody. And especially to say them out loud and to other people because then you have to do something about them. And it's also a good way to make contacts. That's probably how my partner and I came together," says Salzbrunn.

Salzbrunn met her partner socially. She recognized the potential of his idea, and they decided to work together. At the time, Salzbrunn had a part-time job at a small manufacturing company, and she worked nights and weekends on PEAR Systems. The company was not capital intensive, but rather labor intensive. Salzbrunn explains: "When we started in the computer business, we were real pioneers. There were very few companies in the computer field at all. Microcomputers were just starting to happen. We started to build our company very slowly. We didn't even put any money investment into it. It was all our time. The first couple of programs we sold provided the cash that became our working capital."

From this shoestring-operation beginning, the company began

to grow. For the first five or six months they were in business, neither Salzbrunn nor Wilson took any money out. Nearly a year after PEAR began operation, Salzbrunn left her part-time position to work with PEAR full-time. Wilson developed the product, provided technical support, and shared in business decisions. Salzbrunn was responsible for marketing and sales. The company was headquartered at Salzbrunn's home. The reason: "I decided I wanted to start a company that I could run from my house because I had three children, and I was not interested in commuting to New York City [as most of her neighbors did]. I had turned down really attractive offers because of the commute and the hours away." Salzbrunn still prefers to work from her home.

Over the next year, the company continued to grow and acquire a good reputation as a market leader, responsive to the market's demands and problems. And as a woman, Salzbrunn felt easily accepted and respected. "I don't know if it's because it's a new industry, and there were no preconceived old-fashioned prejudices about women, but it's [software and microcomputers] a very good industry for women to be involved with," she says. In addition, Salzbrunn believes women have an advantage in high-tech industries. "I tend to find women are very patient in the computer business. They are much more willing to take the user's point of view than are men, perhaps." (For more on this, see Chapter 2.) In addition, Salzbrunn notes an atmosphere of equal treatment for both sexes: "I don't see job discrimination or women being paid less than men in the same jobs, perhaps because it's such a new business."

In her three-year tenure with PEAR Systems, Salzbrunn worked hard—very hard—and felt pride and accomplishment in her work. Yet she observes, "The start-up of a new business is a lot of work. You find yourself working all day, and all night, and most weekends. And you get tired out. Out of the blue came an offer from Remote Computing Corp. to buy us out. It was a serious offer." After hard thought, Salzbrunn and Wilson negotiated a deal. They traded their entrepreneurial status for cash and more regular employment. "My attitude has always been that it didn't matter what I had to sell. If you're a good salesperson, you can sell anything, as long as it's a good product. So I don't know if there's any real need to have any technical expertise. Certainly because we're in a computer age, it's a comfort to have a good working knowledge of a computer," advises Salzbrunn.

Salzbrunn herself was initially terrified of computers. "I had this horrible feeling that something awful would happen if I touched the keyboard. I really avoided it. We didn't even own a computer. My partner programmed this whole product on borrowed time from other people's computers." Finally they bought a computer (an Apple), but Salzbrunn maintained her distance. When Wilson went away on vacation, Salzbrunn had to use the computer. "Once I did," she remembers, "the fear went away. You realize it's pretty straightforward." Salzbrunn taught herself to be computer literate, but she recognizes that "some people are comforted by taking a class."

What is Salzbrunn's advice to other women? "I think women tend to be very careful and to plan things very carefully. Just jump in. Decide whether you want to be on the marketing end or on the technical end. They're different careers entirely. If you want to be technical, you've got to have technical ability. If you want to be in marketing, either go work for a computer firm—either hardware or software—or start your own.

"There are real opportunities in the micro business because so many people develop products and don't know what to do with them. They have developed good products which go unsold because there's nobody to market them. So let people know you're interested in marketing a product. Get to know people in the computer business, in computer stores. Let people know what you want to do. You may have to sell yourself to be noticed."

Salzbrunn, like many other software entrepreneurs, believes the key to entrepreneurial success lies in smart marketing and sales strategies. (For more on this, see Chapter 8.) Like Harp, Salzbrunn assumed a professional image for her company from the start, a factor that can distinguish a fledgling high-tech company in a young industry like software. Salzbrunn attributes her company's success to these factors as well as its incorporation under the direction of a lawyer and an accountant after a year of operation; hard work; and business and sales skills. Salzbrunn, too, stressed good customer and dealer support, but unlike Harp, dealt with the less capital-intensive field of software.

"Cottage businesses"—small, home-based businesses—are becoming increasingly common in the high-tech world. Computers allow many individuals to work from their homes on various applications, including word processing; writing; editing; computer graphics; architectural, interior, and fashion design; programming; software design; research (using data bases); and even computer hardware design. What Salzbrunn's story reveals is the way cottage businesses may grow. A part-time business can quickly become all-encompassing. That's when an entrepreneur has a decision to make: to take on more of the structure of larger companies, which means hiring employees at all levels, consciously to limit the growth of the enterprise, or to sell. The decision is one of personal choice and of financial and time considerations. It's not always an easy decision to make, and it tends to be irreversible. Often, buy-out contracts spell out the entrepreneur's limits on reentering the market. (Typical would be a clause that says the seller gives up rights to start a competing company or market a competing product for a specified time period, like five years.)

Many entrepreneurs go on to start second and third (and more) companies. The entrepreneurial urge may well be a persistent bug, difficult to ever shake from one's system after the first bite. Do you have entrepreneurial fever? Chapter 8 will help you answer that question. If you do, Lenore Salzbrunn's career illustrates one possible high-tech route to follow. And it's interesting to note that software has called more women to its fold than many other high-tech industries. This might be because of its essentially marketing-oriented nature; many women seem more drawn to marketing than to the technical end of the computer industry.

If you're interested in a home-based business, Salzbrunn's story shows you that a home-based business needn't be small-scale. And if you're interested in developing a successful business as your passport to a senior executive position at a large corporation, Salzbrunn's career demonstrates the viability of that path. In addition, Lenore Salzbrunn is a born networker. She may not belong to formal groups or associations, but she learns from others, and she has happily passed on her hard-won knowledge—and the names of those she has found helpful—to others (time permitting, of course).

"I've found that women are very, very willing to help each other," Salzbrunn says. "Women are much less competitive than cooperative." And while the statement is a controversial one (competitiveness has been considered both a dirty word and a necessary tool for advancement), it's comforting to know that success "happens" to nice people.

AMY WOHL

Amy Wohl parlayed her visibility in the office automation field into a very successful business of her own. On the way, she gained experience in publishing, consulting, and analysis of both the personal computer and office automation markets. In March 1980, Wohl started her company, Advanced Office Concepts Corp., in Bala Cynwood, Pennsylvania. She made use of all of her previous experience by publishing a newsletter and market guides and consulting to end-users. More than four years later, Advanced Office Concepts has become a resource for trade, consumer, and television journalists. Wohl is quoted widely and considered a leading authority on office automation. And she's earned her position.

Wohl recognized that visibility is all-important for a consulting company. Interviews in trade journals and on TV are "worth more than any ad you can buy." Therefore Wohl speaks at numerous conferences, attends computer and office automation shows and conventions, and sends out news releases. All of these efforts have proved more successful than advertising. The drawback is that Wohl must spend 80 percent of her time on the road. How does she manage? "I can buy my way out of a lot of things," she answers simply. Her children are grown and her husband's job doesn't require that he travel. Therefore he can accompany her occasionally on her business trips.

Amy Wohl had already made a name for herself in office automation circles before she started her own company. After earning undergraduate degrees in economics and British civilization from LaSalle University (one of the first women graduates with that major), Wohl wanted to attend law school but couldn't afford it. Instead she began work toward a Ph.D. in labor economics at Temple University. She worked with computers and quantitative analysis there, and after three years—in 1974—she decided to look for full-time work.

Wohl lacked a firm idea of what she was looking for when she came across DataPro, a company which was looking for writers with business backgrounds. They had formerly hired "techies" and tried to teach them how to write, but they had been dissatisfied with the results. "That turned out to be a big bonanza for me," remembers Wohl. "It was like a private graduate school." She was promoted three or four times until she was running all of the office system programs. DataPro became part of McGraw-Hill, and along the way Wohl became executive editor of McGraw-Hill and ran its consulting company.

When Wohl realized she was becoming well enough known to succeed as a full-time consultant—on her own—and earn far more money, she decided entrepreneurship was the next step. Yet, psychologically, it wasn't an easy step to take. Wohl discovered that she was afraid of failure. Before proceeding she had to learn that it was "OK to try what you might not succeed at, as long as you understood what you'd do if you failed.'' Failure, however, has *not* been a problem for Wohl.

Consultancies are ''one of the few businesses where if you start out well-known, you have cash flow.'' (Chapter 8 explains why this is so critical to a company's success.) In addition, Advanced Office Concepts has always had more consulting clients than it could take on, says Wohl. That's an enviable position to be in for any company. Yet publishing can be cyclical, and that's why the editors at Advanced Office Concepts also work as researchers in the consulting division. It's a conscious attempt on Wohl's part to ''stabilize'' the business.

The publishing part of the business grew out of Wohl's desire to create a ''saleable asset.'' (Profit would be realized by selling stock to the public, or selling the company.) There's far more money in consulting, Wohl notes. Personal earnings of $100,000 to $200,000 are not uncommon in the consulting business.

While cash flow has not been a problem, company growth has triggered the demand for large investments of capital to finance required marketing efforts. Wohl turned to venture capital in spring 1983. Venture capitalists invested money for investment purposes only; Wohl gave up no role in management. Second round financing followed a year later.

As a successful high-tech entrepreneur, Wohl has shown herself to be adept at making decisions quickly—with a minimum amount of information—and at risk-taking. And those skills are crucial to high-tech entrepreneurial success. (That's a fact you'll see echoed in case history after case history.) Wohl recognizes that she's had to learn to be aggressive enough to ask for what she wants—or actively seek it—to be successful. Venture capitalists taught her about risk-taking, she says.

As for fear of technology—technophobia—Wohl warns, ''People who are afraid of computing are going to miss out. There's nothing scary about technology. It's under your control. Prove to yourself that you can't hurt a computer in a significant way. And it can't hurt you. It's no scarier than a toaster.''

Amy Wohl's final words of advice: The high-tech world "represents the best opportunity for success in our lifetime, and if a British civilization major can do it, so can you." Inspired? You should be!

SALLY RIDE

Talking about inspiration, Sally Ride may well be the most inspiring high-tech role model for women today. As the first American woman astronaut to go on a space mission, Ride became a media star. In 1983, headline after headline heralded the accomplishments of physicist Ride, whose specialties are astrophysics and laser physics. (Her undergraduate degree is in English and physics; her Master's and Ph.D.—all from Stanford University—are in physics.) And her poise, wit, intelligence, and common sense have won her admirers.

For a person who has been up in space, Sally Ride is admirably down-to-earth. She resents special treatment because of her sex. After she and her fellow astronauts returned to earth, only she was presented with flowers; it wasn't a beauty pageant, and Ride—tired of being seen as the mission's "star"—in a gesture many approved of thoroughly, refused the flowers.

Ride's background gives her something of the appearance of a Renaissance woman, or perhaps—more accurately—an eighties-style Renaissance woman. In college Ride considered joining the professional tennis circuit, but decided she didn't have the discipline to practice tennis eight hours a day. (Discipline is Ride's word; few who know her would doubt that she lacks discipline for anything she wants to do.) Instead of tennis, Ride embarked on a program of running five miles a day, playing intramural sports, and writing and editing articles for women's sports magazines. So how did Ride end up as an astronaut?

At Westlake, a private high school for girls in Los Angeles, Ride learned about the excitement science could offer. The small school's science department consisted of two women Ride thought were terrific. As role models for scientists, they inspired Ride. She decided to major in physics in college, but a course in Shakespeare convinced her to add English courses to her schedule. In her senior year Ride made the difficult decision to pursue physics rather than English. (And at graduate school, she spent about half of her time working with computers)

Four years later—in 1977—Ride was finishing up her doctoral thesis on X-ray astrophysics and looking for a post-doctoral research opportunity in laser physics. By chance she came across an announcement in the Stanford University newspaper (placed by the campus women's center) about NASA's astronaut selection. Ride remembers, "I honestly can't tell you what was going through my mind. I only know I was on my way out of the room to apply while I was still reading the notice in the paper!" Her parents, Ride says, didn't have any preconceived ideas about what their children—both daughters—should do. And while it never occurred to Ride that being an astronaut was a career possibility until she saw the NASA announcement, she has always been drawn to science.

In January 1978, Ride became one of the thirty-five astronaut candidates selected. After a year of training and evaluation (from July 1978 to July 1979) at the Lyndon B. Johnson Space Center in Houston (which included fifteen hours a month on NASA's training airplane and classroom study of the shuttle), Ride became eligible for flight selection. Before she went into training for her flight—in April 1982—she served as ground communicator with the astronauts for the second and third shuttle missions in November 1981 and March 1982. During her space training, Ride spent twelve to sixteen hours per week in the shuttle simulator. In addition, her work at NASA has included satellite deployment, interferon production, and assisting with launch and entry of the space shuttle.

Not part of her written job description, but handled beautifully nonetheless, was interaction with the press. Journalists from around the world wanted to talk to Sally Ride. Some of their questions were silly, and some serious. Ride took them all in stride. After her space mission, she turned down opportunities to make it big on the lecture circuit or write a book about her experiences. Whatever her personal motivations, money is not at the center of them. "I just love my job," Ride says. And Ride gives credit to the women's movement for creating a climate in the late sixties and in the seventies for NASA to accept women. Today approximately 10 percent of the astronaut corps in America are women.

In 1982 Ride married a fellow astronaut. Like many high-tech couples, they share their work—though Steve Hawley's field is astronomy. At thirty-three, Ride has attained one top goal and is no doubt headed for others. Though astronauts are a small, select group, it is encouraging to note that NASA has ended its discriminatory policy against women. Whether or not being an astronaut—or a scientist—interests you, Sally Ride's career illustrates the wealth of high-tech career possibilities available. "Firsts" are often a special breed of people. (See Chapter 7 for a discussion of trailblazers.) They make it easier for others to follow in their footsteps.

The high-tech role models profiled in this chapter are here for many reasons: to inspire you, to show you possible career routes to follow, and to impress you with their stories of accomplishment and success. You may not be interested in being an astronaut or an entrepreneur. You may not even know exactly what you do want to do in the high-tech world. That's where this book can help you.

In the following chapters you'll learn about the careers of other high-tech women. You'll find quizzes to help you discover what would be the right job for you or whether you're cut out for entrepreneurial activity. You'll find detailed strategies for getting training and combatting technophobia, for weighing personal and professional considerations when selecting the right field and position for you, for landing your first high-tech job and moving and advancing. This book offers you assistance in pinpointing the best career direction for you, and it will equip you with top strategies for achieving high-tech success. You'll find out where to go for networking and support, and you'll learn about the head start you have as a woman in high-tech fields.

This book can assist you every step of the way in formulating a high-tech career plan, and in seeing it through. Let the women featured in this book inspire you. And whatever career route you decide to follow, good luck!

2

The Head Start for Women

For years women have suffered in their professional and personal lives from the delusion that they are not mechanically or scientifically inclined or capable. Girls and women have been advised against taking mathematics and science courses and have often been discouraged from approaching computers. Females have been further stymied in high-tech careers by discrimination in the job market and by the need to battle against commonplace misogynistic myths and constantly to prove themselves.

Women have often been accused of not taking their careers as seriously as their male counterparts. Another charge has been that they are not sufficiently competitive or aggressive to succeed in the business world. Some management "experts" have even theorized that since most girls lack the "valuable" background that team sports competition provides, they are forever handicapped as working adults. These "experts" argue that women are at a great disadvantage when they must be team players, as well as when they must compete in office politics.

Women have been advised that they are too social and people-oriented to be happy working with "cold," "unfeeling" machines. They have been warned that programmers are loners, and that the computer industry hungers for math geniuses with a limitless love for number-crunching, *not* social, language-oriented women.

The truth is that the high-tech arena is a place for mavericks —ambitious individuals able to buck the status quo and ride the rocky road of change. Technical skills, business skills, *and* interpersonal skills are all crucial to success here. And while the stereotypical high-tech success story is male, the reality (as shown in Chapter 1, and further demonstrated in Chapters 4, 7, and 8) is not.

Women *do* succeed in the computer industry and in a full range of computer-related occupations. Of course, the reasons for their successes are as individual as the women themselves. This chapter highlights the most pertinent research available to indicate why women as a group *tend* to achieve in high-tech fields. This is not, of course, to imply that every woman is capable of succeeding in each high-tech position. (See Chapter 5 for help in finding the right job for you.) Nor is it my purpose to avow that every woman holds qualifications superior to every man in high-tech industries. (Reverse discrimination is no prettier than any other kind of discrimination.) Moreover, these pages do not slight the achievement of any successful high-tech woman.

So why this chapter? Because women—and men—need to be reminded that women, by virtue of their upbringing, training, and traits that may be physiologically based, can do well in high-tech careers. It's a statement that should be accepted as a given. Since frequently it isn't, the supporting evidence may be found in this chapter. Women are still informed too often that they belong in the home or that they don't require the same advancement or financial opportunities as men. In a climate where such misconceptions are still common, being equal may not always be sufficient.

Here I'd like to point out some highly significant factors that can contribute to women's high-tech successes. For any woman who has feared that she was disadvantaged in this career category, here's enlightenment and encouragement. And for every woman—or man—who has suspected feminine advantage or equality in high-tech areas, here's an explanation.

Different skills and personal attributes are required in individual high-tech positions (as will be illustrated in Chapter 5) so let's examine them only briefly here. Programming, contrary to popular belief, is not principally related to math expertise. What counts far more, say industry experts, are such traits as attention to detail, concentration, patience, ability to work under pressure, capacity to work alone and with others, adaptability, commitment, persistence, and hard work. There is evidence that these traits are more likely to be found in women than in men.

ATTENTION TO DETAIL

Psychological studies show a female superiority in attention paid to detail. One study was performed by the well-respected Johnson O'Connor Research Foundation (a non-profit educational organization which studies human abilities and advises men and women in education and vocational decisions).

Why might women be better at paying attention to details? The answer is uncertain. Most likely it's due to the training girls receive in childhood. Girls are taught to be neat. Sloppiness is tolerated less in girls than in boys. As they grow up and learn to balance a concern for details with a concern for others, females tend to be less likely to leave loose ends or unchecked work.

In programming, details are everything in creating a flow chart and writing code. Details must be scrupulously manipulated to develop an effective program. And debugging—checking the program for errors—is an integral part of the programmer's job. A programmer who is careless about details will find that the costs of sloppiness are high in time and effort—and job success or failure. A good eye for detail is also required in other computer occupations: design, analysis, repair, control, operation, as well as in management.

Esther Dyson, president of EDventure Holdings, Inc. in New York, says, "Women are often educated to pay more attention to detail, to be more painstaking. . . . To that extent they should be better at programming. That's not biology, that's training."

Commodore Grace Hopper of the U.S. Navy observes that

when it comes to programmers, "I prefer to have women. They have, innately, a tendency to finish things—to document a program completely and wrap it up in a nice package." Hopper thinks this tendency comes from "our backgrounds."

"Women do a much more organized job," agrees Genevieve Cerf. She has taught computer science and now teaches electrical engineering at Columbia University. "They write neat programs, beautifully laid out. Women are much more verbal in their descriptions; variable names are well chosen. They're terrifically logical. They don't do crazy things. Male students I have do insane things that can't possibly work, and I've rarely seen women do those things."

Organization and attention to detail don't only pay off for programmers. "I know several women who are assistant managers—people who oversee the running of an entire computer system—and they do a terrific job," says Cerf, "because they're more interested in keeping things organized and well-maintained. The men like that because they don't like to worry about details. Men are often a lot sloppier about how they maintain things," she claims.

LANGUAGE SKILLS

A number of studies reveal women's superior linguistic skills. What does this mean, and how does it count in high-tech careers? Women have a greater aptitude for acquiring languages and professional terminology. This was the conclusion of the previously mentioned Johnson O'Connor study. Computer languages, whether they are for business or scientific applications—COBOL, BASIC, LISP, PASCAL, ADA, PL/1, FORTRAN, ALGOL, or others—are more easily learned by those with linguistic aptitude. And the more easily and quickly a language is mastered, the more valuable an employee a programmer is.

Similarly, women rate higher than men at verbalizing ideas, according to the Johnson O'Connor study. This ability can make women more effective managers, teachers, and salespeople and can make them successful in marketing, promotion, advertising, and technical writing. Women are considered more sensitive to verbal style, which can result in heightened persuasiveness, as well as the ability to tailor text or speech to particular audiences or individual needs.

Studies by Yale psychologist Faye Crosby, Ph.D., indicate that women tend to give more feedback in conversations than men. This trait is an important attribute for successful, growing, upwardly mobile managers. The feedback may include one word responses like "Yes?" or "Then?" or non-verbal signs like smiles and nods. The more feedback from a listener, the more comprehensive the speaker's narrative and the greater the listener's comprehension. So conclude Cornell University psychologist Robert E. Kraut, Ph.D., and colleagues from recent research. Women also tend to look into a speaker's eyes while listening, says Crosby. This is desired and appreciated by management and staff alike. It is interpreted as intense interest. And women are also more likely to be more sensitive to context. Crosby notes that the traditionally female communication style is now being emulated by men.

PEOPLE-ORIENTATION

The very trait often used to discourage women from entering high-tech fields is actually an advantage. Women do tend to be more oriented toward other people than men are. How can this be an advantage in a field where the stereotypical hacker is a young male loner, addicted to computers? The answer is that hackers are the exception, not the rule, when it comes to *career* success stories. Hackers are typically students. After graduation (if they ever graduate), hackers generally must learn to fit into normal working conditions or make their own way. As entrepreneurs, hackers may have the intellectual and personal drive necessary to succeed. However, they tend to be sorely lacking in an understanding of their customers or clients and deficient in down-to-earth business sense. If you're not a hacker, rejoice.

You can enjoy working with computers and other high-tech equipment and still enjoy being with people. It's a rare high-tech employee or manager who needn't interact with others. If your goal is management or starting your own company, interpersonal skills can make or break your career. You'll need to get along with others, accept criticism and mete it out when necessary, know how to listen and communicate, know how to delegate responsibility, evaluate the work of others (and of yourself), be a team player, and think in terms of the user or client.

"User-friendly" software—which is easy for the consumer to use and the trend of the future—is a direct result of people-orientation. Well-written documentation, smartly directed advertising, on-target proposals and memos, and products designed to meet consumer needs all pay off on the bottom line. Thinking in terms of the user is not simple courtesy, it's a professional necessity. And the hacker who may be a technical whiz kid but cannot satisfy the consumer's needs offers no competition in the marketplace.

"Women are more interpersonally focussed than men," says Dr. Milton Kramer, a psychiatrist at the University of Mississippi Medical School. And this can make them better team workers, as well as better managers. Interacting well with others is a key attribute of a successful programmer functioning well on a programming team *and* of an effective customer support representative, systems analyst, computer science teacher, computer salesperson, technical writer, and consultant—in short, anywhere where an individual is better served by thinking in terms of the user, instead of being restricted to her or his own point of view.

According to Gay Bryant, former editor of *Working Woman,* a 1983 study undertaken by the magazine revealed that "more than men, women view themselves as an integral part of the workplace, and they think of themselves more as a part of a team." In the close, collegial atmosphere of many a small, growing, or start-up high-tech company, feeling part of the team is as important to employers as many a specific technical skill. Where spin-off companies are as common as sunny days in Silicon Valley, team spirit (sometimes known as company loyalty) is valued.

People-orientation also plays a part in office and corporate politics. The manager who takes an afternoon off to catch a son or daughter in a school play is less and less an anomaly. As offices become more humanized, life enters the office. In young high-tech companies, managers are recognizing that satisfied, happy employees are better employees. And women have helped spearhead this change. The day I spoke to Helga Johnson, president of Tago, Inc., she was answering the telephones herself. Why? She had sent her employees home to prepare as they wished for Christmas, only days away. That's not typical behavior for the president of a biotechnology company that grossed approximately $1.82 million in 1983, but it doesn't seem to be hurting her professional image—or employee satisfaction either.

A humanized work atmosphere also pays off in reduced stress and stress-related diseases like migraines, ulcers, colitis, and heart disease. Healthier workers save a company money on reduced sick leave, early retirement, and lower replacement costs. (Women are less injury-prone on the job as well, according to a 1983 study of the Center for Disease Control.)

VISIBILITY

In male-dominated fields, a significant advantage women have is visibility. Like many other advantages, visibility can also be turned to one's disadvantage. Women computer salespeople and high-tech entrepreneurs are more visible and thus more memorable because of their minority status. And any sales representative must be concerned with presenting an excellent pitch for her product *and* being remembered.

A business owner requires company recognition from potential customers or clients to succeed. If an owner stands out because of her sex, this can facilitate getting publicity for the business.

Is high-level recognition by virtue of gender ever a disadvantage? Only if one's job or performance—or company—is not up to par, or if one's sex is seized upon as a reason for harassment or discrimination. Bestowing sexual favors for career advancement is growing exceedingly rare in today's business world. Unfortunately, sexual harassment and discrimination are not as rare. Yet there are laws, rules, and programs designed to help women (and minorities) fight harassment and discrimination. If you're willing to stretch the category of female advantages, you might be inclined to include these programs under the umbrella. Read on for what you should know about Title VII of the Civil Rights Act, the Women's Educational Equity Act, other protective legislation, affirmative action programs, as well as scholarships just for women. (Investigate the American Association of University Women—AAUW, American Women in Science—AWIS, Society of Women Engineers, women's networks and associations, Clairol, Inc., etc.)

Programs that encourage women to enter science and high-tech fields are offered by the Math & Science Education Program for Women at the Lawrence Hall of Science in Berkeley, the Math/Science Network (based at Mills College in Oakland), the Girl Scouts, Computer Town USA!, Mathematical Association of America, AAUW, and AWIS.

THE LAW ON YOUR SIDE

The Equal Pay Act of 1963 was an amendment to the Fair Labor Standards Act. It went into effect in June 1964, and is administered by the Wage and Hours Division of the Department of Labor. The law prohibits employers from discriminating against either sex in pay for jobs of similar skill, effort, and responsibility, performed under similar working conditions. This act covers employers and unions only in industries involved in interstate commerce.

Title VII of the Civil Rights Act of 1964, as amended in 1972, led to the development of the Equal Employment Opportunity Commission (EEOC). This act prohibits discrimination on the basis of race, color, religion, sex, or national origin in hiring, classifying, training, promoting, and other conditions of employment. Advertisements indicating discriminatory preferences ("Man wanted") were made illegal by this act. However, Title VII applies only to employers of fifteen or more workers, employment agencies, labor organizations, and labor-management apprenticeship committees.

The Age Discrimination in Employment Act of 1967, which became effective in June 1968, bans discrimination against individuals aged forty to sixty-five in all work conditions, from hiring to compensation. This legislation, administered by the Secretary of Labor, pertains to employers with twenty-five or more workers, as well as to employment agencies and labor organizations.

In addition, legislation exists to promote affirmative action in hiring and to ban discrimination in educational programs. Executive Order 11246 (1965), amended in 1968 (Executive Order 11375), covers employers who have federal contracts or subcontracts valued at more than $10,000. These employers are not only prohibited from discrimination; they are also required to take affirmative action to insure equal opportunity. In all advertising, these employers must state that they are affirmative action/equal employment opportunity employers. Revised Order 4 requires employers with more than $50,000 in federal contracts, or fifty or more employees, to file their affirmative action plans with the Office of Federal Contract Compliance. These plans must include both goals and timetables.

Title IX of the Education Amendments Act of 1972 became effective in July 1975. It prohibits discrimination on the basis of sex in all federally assisted education programs. This law covers public and private preschools, elementary and secondary schools, vocational and professional schools, public undergraduate schools, and graduate schools. Title IX is administered by the Department of Labor.

ENCOURAGING PROGRAMS

The myths and stereotypes that are problematical for women in pursuing high-tech careers start early on. Young children learn from their parents, teachers, and each other, and are quick to pick up spoken or unspoken cues about "appropriate goals." Too frequently, boys dominate in computer literacy classes and in computer use. Often parents purchase computers for their sons, and ignore their daughters' equal need for computer time and stimulation.

These are some of the underlying reasons for the development of programs like EQUALS at the Lawrence Hall of Science in Berkeley, California. EQUALS encourages girls and young women "to become interested in and proficient with computers." How? EQUALS is a *teacher* education program. Teachers, counselors, and administrators at all grade levels are taught about why the issue of computer education for females is so important, how to provide instruction and materials which encourage confidence and competence in using computers, and how to create student motivation toward computer-related occupations by means of role models and special activities. EQUALS provides statistics about male and female computer class enrollment, the use of computers in occupations, salary ranges for these occupations, and opportunities for women in these fields. The message EQUALS puts across is clear: Females need to be able to use computers; their careers and finances depend on it.

Other programs designed for women include the Math/Science Network's conferences, "Expanding Your Horizons" (EYH), through Mills College in Oakland, California. These conferences are aimed at seventh through twelfth grade female students. The Network, made up of 1,000 scientists, engineers, educators, and others, works toward increasing the participation of women in math and science. Similar workshops are sponsored by the Mathematical Association of America. American Women in Science (AWIS), and the American Association of University Women (AAUW). The AAUW offers scholarship assistance to women returning to school, advancing their careers, or changing career direction through a program called Project Renew. AAUW's Future Fund trains trainers in computer literacy who then train other AAUW members at conventions. (AAUW is headquartered in Washington, D.C.)

FEMININE INTUITION

Is feminine intuition myth or reality? And more to the point, is intuition pertinent to high-tech career success? Answers: Feminine intuition is no myth, and intuition plays a remarkable role in *any* career (and social) success. What is intuition? It's a faculty of knowing something directly without the use of rational processes, according to Philip Goldberg. Goldberg is author of *The Intuitive Edge*. "We can't trace any steps that went into the knowledge," he adds.

Intuition has been undervalued in Western culture, yet it's an incredibly useful faculty in interpersonal relationships at work and in social settings. While everyone may have the ability to be intuitive, women tend to be more intuitive and more skilled at using their intuitions. Some possible explanations:

- "Anyone who's in a position of less power has to be aware" of non-verbal cues, says psychologist Ulrich Neisser, Ph.D., of Emory University. "I think that it's true that in general oppressed classes are more sensitive. . . . It's not that it's in the genes," Neisser adds. This sensitivity is part of what makes up intuitive ability.
- "Women do seem to be better at processing emotional information. They are better at understanding non-verbal information

and reading facial expressions and are more sensitive to slight variations in sound and color,'' observes Philip Goldberg. He suggests cultural training or environmental factors hold the key to this situation.

• Social conditioning is the explanation of psychologist Frances Vaughn, Ph.D., author of *Awakening Your Intuition.* ''In our society men are trained to repress their feelings more,'' she says. ''This act of control inhibits the role intuition is allowed to play in one's life. Men are more likely to distrust their intuited thoughts and not to test them. Instead they will dismiss them and overstress the rational,'' claims Vaughn.

• There may be a physiological basis for female superiority in intuitive ability as well. Brain lateralization—the way the two hemispheres of the brain are connected—may provide an explanation, notes neuropsychologist Jerre Levy, Ph.D. This University of Chicago professor pioneered the study of brain lateralization, and she discovered that female brains tend to be less lateralized, or drawn apart, than male brains. This allows the more closely connected hemispheres of the female brain to communicate more quickly. And this enables women to incorporate and assess more details and nuances than men. That's one way of defining intuition physiologically.

• A psychological explanation for female intuition can be found under the heading of ''incidental learning.'' In studies of the incidental details males and females pick up in the course of concentrating on other information, women always score far higher. Incidental information can include facial expressions, colors, sizes, shapes, sounds, and more. This is often what makes up intuition.

• Studies of young children show that infant girls are more sensitive to sounds and to changes in tone and in voice inflection. Girls focus more on faces than on things. And as adults, women are better skilled at understanding body language, deciphering facial expressions and tone of voice. These skills are critical to picking up cues about other people, the stuff of so-called women's intuition.

OK, where does all this evidence leave us? Convinced that women tend to be more intuitive then men, but what does this mean when it comes to high-tech career success? Intuition isn't only useful in personal relationships. In every career, people play a crucial part—as bosses, co-workers, subordinates, mentors, customers or clients, loan officers or venture capitalists, references, and more. The individual who is skilled at assessing others quickly and accurately—as helpers or detractors, as individuals who need to be shown vs. individuals who need only to be told, as hard workers or lazy thinkers—has an edge in the marketplace. For intuitions like these are at the very heart of what is needed for business and career success.

Women are skilled at intuitions regarding equipment design and use, advertising and marketing angles, sales pitches, financial decisions, and more—*not* just regarding people. None of this is to say that intuitions are always correct or should always be trusted. Every intuition worthy of a second thought should be examined and *tested*. Just as a scientist begins with a hypothesis and then proceeds to test it, a person should check out an intuition, not act on it impulsively.

Many successful entrepreneurs are highly intuitive. To accomplish all that entrepreneurs must to survive, energy and drive are often insufficient. A finely honed intuitive sense often provides the extra ingredient for success.

Men may not be lacking in intuition, but their intuitions—by virtue of upbringing or cultural expectations or even physiology —tend to be of a different sort. The advantages women enjoy in high-tech careers may well come down to an easy acceptance of their intuitions. By practicing and cultivating their intuitions, at work and after working hours, women are fine-tuning a prime business advantage.

PROBLEM-SOLVING ABILITY

Problem-solving is integral to many high-tech positions, from programming to systems analysis, engineering to data base management, technical writing to marketing. And it turns out that recent studies indicate that women solve problems very differently from men. In fact, the method women typically use—focusing on the particulars of the case rather than seeking a general rule that applies—is the very same method a *Harvard Business Review* article applauded as the method of "managers who consistently accomplish a lot . . . ," according to business professors Wickham Skinner, Ph.D., and W. Earl Sasser, Ph.D.

Carol Gilligan, Ph.D., author of *In a Different Voice,* studied the way women solve problems and reach judgments or answers. Her findings, that women are inconsistent in their manner of attacking problems, fits in perfectly with Skinner and Sasser's evaluation of the optimal problem-solving method for managers. Thus another trait formerly viewed as a feminine disadvantage can now more rightly be acclaimed an advantage.

In addition, psychological studies have shown that women are better skilled at screening out irrelevancies and working under pressure. Many high-tech problems are "emergencies"; under tight deadlines a solution must be developed and implemented. A systems analyst must prepare a cost-effective proposal for a company's automation, knowing she is competing with a rival consulting firm. A computer scientist at work on a new hard disk drive can't afford to panic at the least sign of a setback. Similarly, an entrepreneur whose company's future depends on the timing and success of a new, ambitious word processing software package must be able to balance drive with the calm needed to weather the storm. Perhaps this ability is related to an individual's ability to balance cooking dinner, listening to a child's tale of her day at school, writing a shopping list, and sharing a joke with a spouse. Men can do this too, but women tend to be better.

Perhaps it's more years of practice. Whatever the origins, it's a trait that gives women an advantage in the high-tech marketplace.

CAREER COMMITMENT

Career and company commitment—as distinguished from workaholism or blind loyalty—are traits high-tech employers value highly. Robert Half, president and founder of Robert Half of New York, a large professional data processing recruiter, has concluded from a study of job orders filled by his company in the accounting field (at salaries between $15,000 and $50,000), that women got the job 73 percent of the time. Why? Probably because women were more frequently seen as loyal to the company. (For more on the significance of company loyalty, see Chapter 7.)

Another significant shift is the way women are perceived in the marketplace today as being more committed to their work. A 1982 survey conducted by the Public Agenda Foundation (a New York non-profit organization headed by Daniel Yankelovich) for *Working Woman* magazine revealed that 71 percent of the women exhibit strong commitment and self-sacrifice regarding their work, compared to 64 percent of the men. In addition, women believe in the work ethic more strongly than men: 56 percent of women vs. 48 percent of men. And this study was not alone in its conclusions.

A 1984 study by the American Management Association of its members (nine times as many males as females) reported that female managers are both more career-oriented than their male counterparts and that they derive more satisfaction from their jobs. The study was conducted by researchers Warren Schmidt, Ph.D., of the University of Southern California and Barry Posner, Ph.D., of the University of Santa Clara. Posner conducted a larger study with colleague Gary Powell, Ph.D., of the University of Connecticut and reached the same conclusions.

"Women seem to be much more committed than men," Powell says. "These results may convince companies that women actually have an *advantage* as managers."

Researcher Patricia Kosinar points out that an explanation may be found in the way women are raised. Girls are taught that following directions and meeting expectations are what they should do. These have been two major elements of the traditional female role. And this may account for women's higher work ethic.

While commitment and work ethic and company loyalty are desirable traits for an employee to have, employers look for more in high-tech managers. And in women they find another six advantages that tend to make women better suited for management, according to tests of aptitudes performed by the Johnson O'Connor Research Foundation. The aptitudes where women excel include:

- word association (useful in acquiring new languages or professional terminology; a must for most high-tech employees)
- observation of details
- abstract visualization (ability to deal with abstract problems, ideas, and principles; an asset in management, writing, and other non-technical occupations)
- "ideaphoria" (the rate of the flow of ideas used in activities that require persuasion and verbal fluency; crucial to careers in sales, teaching, writing, or advertising)
- finger dexterity (which may be useful in using keyboards or tools in repairing, designing, or operating high-tech equipment, or in good old "paper shuffling")

RIGHT BRAIN/LEFT BRAIN

You've probably seen many articles in recent years about the differences between "right brain" and "left brain" people. Popular books on unleashing your creativity or drawing from one side of your brain use as their premise research regarding the different strengths associated with each hemisphere of the brain. To be precise, these writers are really talking about the right and left hemispheres of the brain. Most individuals are governed more strongly by one side than by the other. For women, left brain superiority is most common. And the two hemispheres tend to communicate more in women. What does all this mean?

Articles in popular magazines translate these physiological facts into the following conclusions:

- Women are more verbal and articulate.
- Women are more sensitive to touch, taste, smell, and sound.
- Women are better at reading comprehension.
- Women are better at talking about their emotions.
- Women are better at reading facial expressions.

Are these statements true? Not always.

The field of brain research is one rife with controversy. Any statement made by a researcher may be taken up by the press and either exaggerated or simplified. And it's a small step from a scientist's view of "tendency" to a reporter's statement regarding actuality or destiny. Brain research regarding sex differences demands close attention.

Men and women are different. How different and what those differences mean are what's at issue now. Most brain researchers are careful to couch their pronouncements in words like "tendency" and "likelihood" and "relatively." As neuropsychologist Jerre Levy, Ph.D., notes, "Just because one hemisphere is superior, it doesn't mean that the other is incompetent." Therefore to say that women tend to be exceptionally skilled at language tasks is not to say that men are inarticulate or illiterate. Nor should we conclude from a male tendency to do better at maps and mathematics and spatial reasoning that women are incompetent at these tasks.

Why then should we pay any attention to brain differences at all? Because they indicate tendencies that can be encouraged or discouraged. Women are often encouraged to talk about their feelings. What has this got to do with high-tech careers? Talking about feelings—about convictions regarding a new product, about hunches concerning a group of job candidates, about intuitions that have to do with marketing plans or financing decisions —gives women a language that can be very persuasive and very effective (and impressive) in high-tech business meetings. And if you compound an initial tendency with societal expectations and years of practice, you tend to come up with a decided advantage.

Psychologist Eleanor Emmons Maccoby, Ph.D., of Stanford University notes that girls' thinking as compared to boys' is ". . . more global, and more persevering." Being able to look at the larger picture and persist counts for a great deal in many high-tech occupations, like programming, systems analysis, engineering, data base management, executive recruiting, management, sales, marketing, and more.

Dr. Richard Restak, neurologist and author of *The Brain: The Last Frontier,* believes that women have a tendency to be more proficient at linear logic skills. These skills, again, are at a premium in programming, systems analysis, and engineering.

All this data can only support the following conclusion: while women may have certain advantages *as a group* because of left hemisphere superiority or brain lateralization (the way the two hemispheres are connected), these advantages cannot predict how any one woman will do at a specific job or task. The point of this section is not to suggest that all women have an overwhelming advantage over men, but to remind women and men that solid research exists to counteract traditional stereotypes of what each sex can—and can't—do. In a world where women are still told they can't succeed at many high-tech careers because "they aren't mathematically or scientifically inclined," this data helps balance the scales.

FEMALE ROLE MODELS

Chapter 1 introduced you to many female high-tech role models. How can these women be considered an advantage for you? Think about what motivates you to succeed and what helps support you on your path to career success. Chances are that knowing what you want to do is possible helps immensely; similarly, recognizing that it has been done by other women adds encouragement and support.

It's difficult to be a "token woman," and it's difficult to be a "first" as well. Many ambitious women (and men too, today) worry about balancing career success with a happy personal life —a good marriage, healthy and happy children, and some private time too. The model of feminine career success not too long ago seemed to preclude marriage and family. And many younger career women today still insist that you can't have it all. Role models show a way to balance your priorities. You can choose to make your business your "baby," or postpone children until your career is on an even keel. Or you can postpone your career advancement until your children are in school, or choose to handle it all at the same time as best as you can. Female role models can show you how they did it. The lessons you learn are up to you.

Female role models who have achieved success in high-tech careers can also point out possible career pitfalls and setbacks in advance. It can be like having a mentor or sponsor without the personal relationship. Only recently, the major business schools of the United States, such as Harvard Business School, have begun to include women business owners and managers in their case studies. Learning by example is one of the more effective ways to learn. And when you can identify with a role model or example, so much the better.

CLIMATE OF CHANGE

Women intent on high-tech career success today have a distinct advantage over their older sisters or mothers. Affirmative action programs are only one sign of the change in times. Despite the long list of myths and stereotypes that hinder women's career progress, high-tech companies tend to be more interested in women employees than ever before. Why? Let's sum up twenty reasons explored in this chapter and elsewhere in this book:

1. Women tend to be more verbal and articulate, mastering languages and professional terminology faster.
2. Women tend to put themselves in the shoes of users, clients, and others.
3. Women tend to pay attention to details.
4. Women tend to be more people-oriented.
5. Women tend to have a greater investment in the work ethic.
6. Women demonstrate greater company loyalty.
7. Women managers tend to be more satisfied by their jobs.
8. Women tend to be more intuitive.
9. Women tend to be more creative about problem-solving.
10. Women tend to work better under pressure.
11. Women seem to be better skilled at reading facial cues, body language, and other non-verbal communication.
12. Women tend to be more persistent.
13. Women tend to be more sensitive and accessible.
14. Women tend to be good team players.
15. Women are less injury-prone.
16. Women are seen as more nurturing.
17. Women tend to process information faster.
18. Women tend to be better listeners.
19. Women tend to excel at linear logic.
20. Women's fine motor skills are superior.

Not an unimpressive list, don't you think? Add to the above list the fact that high-tech companies tend to number among their executives younger, more liberal men than more conservative smokestack industries (like steel), and you see the support women enjoy today in high-tech careers.

"A lot of the new high-management people are young men," agrees Walter Kristiblas, who teaches computer science in New York. "And I think their attitudes toward women in general are radically different from those of the typical male manager that one encountered twenty or twenty-five years ago." This doesn't guarantee your success, but it sure makes it more possible.

Additional help is available in the form of hundreds of women's support groups and networks for women in computer-related occupations, in mathematics and science, in teaching and sales, in marketing and advertising, as well as in high schools, colleges, and graduate schools. These groups are both professionally oriented and socially oriented. They provide workshops on equipment, technical and business techniques and skills, job banks, collegiality, and networking opportunities. They help create new "old girl networks" to counterbalance the advantage men have enjoyed over the years with their "old boys' networks." These groups provide more than membership cards. They can lead to career advancement, promote professional visibility, and further technical and business education. And ten years ago, most of these organizations did not exist. (Many of them owe their existence to the consciousness-raising or "c-r" groups of the seventies, which were designed to offer personal support.)

"They're creating a communications link with one another. I think before long you're going to see . . . women having an advantage when a position becomes available," says Walter Kristiblas, who teaches in New York State's Board of Cooperative Educational Services (BOCES).

When advice on business and career decisions is what's needed, organizations exist to provide this for women as well.

For names and addresses of all these helpful organizations and societies, see Appendix I.

The bottom line is, of course, that you must succeed on your own merits. But isn't it nice to know there are many good reasons why you, as a woman, can succeed?

3

Technophobia and Training

What is technophobia? Who suffers from it? Can it be cured? How? Before moving on to the answers to these questions, take this test to assess your TP Quotient (how technophobic you are). Choose the answer that you feel most accurately expresses your feelings.

Test Your TP Quotient

1. When I see a computer, I:
   (a) start trembling
   (b) get sweaty palms
   (c) freeze
   (d) feel slightly uneasy
   (e) feel relaxed or challenged

2. When I touch a computer keyboard, I:
   (a) remember how much I hate to type
   (b) wish the computer was screen-operated
   (c) feel immediately comfortable
   (d) feel confident
   (e) feel like a secretary and hate it

3. When I read about new
   (a) feel optimistic

advances in technology, I:
(b) feel frightened about the future
(c) wonder how they did it
(d) wish I could do that
(e) get bored

4. I would rather learn to program a computer than:
(a) see a live rat
(b) be touched by a live rat
(c) see a dead rat
(d) eat ice cream
(e) be stuck in a dead-end job

5. The worst job that I can imagine is that of:
(a) garbage collector
(b) junior high school teacher
(c) hazardous waste material guard
(d) computer programmer
(e) systems analyst

6. Guaranteed to make me laugh is:
(a) the idea that I can easily understand high-technology
(b) the suggestion that I could be a math whiz
(c) Doonesbury cartoons
(d) Joan Rivers
(e) my bank account

7. Trying to make sense out of the typical software manual is so frustrating that:
(a) I give up in tears
(b) I call the store for help
(c) I need plenty of time
(d) it's no picnic
(e) I never try

8. One of my greatest satisfactions is:
(a) being right
(b) being right and having everybody know it

(c) being wrong and having nobody ever find out
(d) that I don't have to drive a car or use a computer
(e) taking something mechanical apart and putting it back together again

9. One of my more significant motivating forces is:
   (a) fear
   (b) guilt
   (c) curiosity
   (d) desire to achieve pleasure
   (e) none of the above

10. I think the word that best describes high-tech entrepreneurs is:
    (a) incomprehensible
    (b) risk-taking
    (c) creative
    (d) success-oriented
    (e) indomitable

OK, go back to tally up your scores as follows:

1.	High: a,b,c,d,	Low: e
2.	High: b	Low: a,c,d,e
3.	High: b	Low: a,c,d,e
4.	High: a,b,c,	Low: d,e
5.	High: d,e	Low: a,b,c
6.	High: a,b	Low: c,d,e
7.	High: a,e	Low: b,c,d
8.	High: c,d	Low: a,b,e
9.	High: a	Low: b,c,d,e
10.	High: a	Low: b,c,d,e

Give yourself ten points for each "High" answer, score zero points for each "Low" answer.

Key

A score of zero indicates you're free of technophobia. Congratulations!

A score of 10 to 30 indicates you're only slightly fearful of technology. Reading this book should help you destroy the last vestiges of computer fear you may feel.

A score of 40 to 60 indicates that computer fear may be holding you back. Read this chapter carefully to find out how you can overcome it.

A score of 70 to 100 reveals that computer fear is a real problem for you. Try this quiz again after reading this chapter and following its suggestions. If your score doesn't improve considerably, high-tech fields may be too threatening for you.

How did you score? If your TP Quotient is high, use this chapter as a guide to conquering your fear. If you believe you're not technophobic at all, you may want to skim the first part of this chapter and then move on to the section about computer training and education.

WHAT IS TECHNOPHOBIA?

Technophobia, also called computerphobia, cyberphobia, and a host of other names, is a major problem for individuals and businesses. To put it simply, it's fear. Fear of failure. Fear of the unknown or of the unfamiliar. Fear of appearing stupid. Fear of breaking the machine. Fear of being replaced by a computer. Fear of losing one's privacy or being watched by Big Brother, a la George Orwell's classic *1984*. Fear of change. Fear of becoming dependent on a computer or even addicted. Fear of being left out or left behind. Fear of loss of control. Fear of math and of anything mechanical. As you can see, it's a complex fear.

"People are afraid of the new technologies," says Edward Cornish, president of the World Future Society. "They don't understand what's going on and don't even know what questions to ask." Technophobia is a result of trying to cope with an ever changing world. The changes are occurring too rapidly for many to accept or understand.

Does age figure into the fear? "The older the person is, the worse it is," says Arnold S. Kahn of the American Psychological Association. Dr. Mary Gray, professor of mathematics at American University, agrees. "In people over thirty-five it occurs more frequently."

Fear of technology can be expressed in feelings of "anxiety, feelings of rootlessness, and for some a sense that even their work identity is endangered," says Dr. Herbert Freudenberger, a New York psychoanalyst and author best known for his work on burn-out.

A fear becomes a phobia when it is intense enough to interfere with one's life. Technophobia can keep you from pursuing a satisfying career or just keep you from feeling comfortable in your interaction with bank automatic teller machines, airline reservation clerks, and the like. Today there are countless technophobia seminars and workshops and books and articles nationwide. What is best for you?

First, examine how your fear makes you feel. Do you think you're missing out because of your technophobia? Do you believe your career is suffering? Do you suffer unnecessary anxiety? Do you feel resentful of technically proficient individuals? If you've answered yes to any of these questions, you're probably ripe for some positive thinking—and some computer training.

Computer literacy—being able to use computers—is not a particularly difficult skill to acquire. While there is a mystique surrounding computers still (walk into any computer store and try to understand the unfamiliar terminology thrown around by most salespeople!), it's easy enough to lift the curtain. Learning about computers involves learning some new terms and techniques. And this knowledge is at the heart of combating technophobia.

"Computers have the capability to act as tools for us everyday folks, as *knowledge amplifiers,*" says psychologist Dr. Donald A. Norman. "This means they should help us out with everyday tasks, making it possible to do things we could not do before. Computers are tools and should be treated as such; they are simple tools in the same category as the printing press, the automobile, and the telephone," adds the University of California at San Diego professor.

FEAR OF THE UNKNOWN

Consider the basic nature of technophobia. It's fear of the unknown. Remember the first time you learned how to ride a bicycle or ski downhill or dive from the high board or even take a test. You probably felt fear, whether the emotion was expressed in sweaty palms, trembling, stomach upset, or even high blood pressure. Most people respond to new things this way, so you are not alone. And yet most people learn to ride bikes, drive cars, and take tests. (And if they really want to, they can learn to ski or dive, maybe not with Olympic caliber, but well enough to enjoy the sport—or prove something to themselves.)

Fear of the unknown is probably the largest part of technophobia, says Betty Vetter of the Scientific Manpower Commission in Washington, D.C. And this can be complicated by being told by other people that you can't do something, adds Vetter.

If you're a fighter by nature, bear in mind what some men—and women—would have you believe: that women are innately inferior when it comes to mathematical or scientific skills. It's not true. While math and science may not have been your best subjects in school, there's no reason you can't learn how to use a computer or enjoy a high-tech career. Believe me!

Math educator Sheila Tobias is famous for her groundbreaking book, *Math Anxiety*. It's a fear women may feel more than men, probably because they haven't been encouraged as much as men have to succeed at math. Both Betty Vetter and Dr. Mary Gray agree that many women experience math anxiety. And many women relate computers to math, notes Gray.

You do need "a good math background" and some logic to really do something with computers other than just use them or punch in data, says Gray. Vetter points to some alarming statistics about the participation of women in math courses from high school on—and the connection between the low numbers in classes and the low numbers of women in science and engineering professions, and in high-paying jobs in general:

If a girl doesn't take math in high school, she "knocks herself out of being eligible for three-fourths of all professional level occupations, three-fourths of all college programs, and all the high-paying jobs she might want," observes Vetter. That means that today only one out of a thousand women has a Ph.D. in a quantitative field, like engineering or mathematics, compared to five out of a thousand men.

Is math anxiety surmountable? Yes! Dr. Marilyn Berman, assistant dean of the School of Engineering at the University of Maryland, counts herself among the sufferers. (Her Ph.D. is in administration, not engineering. She is the first person in her job *not* to have an engineering degree.) Berman counsels fellow sufferers to take programs that confront math anxiety head-on. Taking a low-level math course and succeeding at it can build confidence, she observes. Then an individual can move up to more difficult courses. Find someone who is sympathetic as a teacher, advises mathematician Dr. Gray. And seek out schools or employers where many women hold technical positions, she adds. Role models can help convince you that you can do it too.

Exactly what relation does math have to using computers? This depends on whether computer use involves simple computer literacy or more complicated programming or systems analysis techniques or even engineering or designing skills. Using or operating a computer today requires very little mathematical knowledge. In fact, if math is a source of frustration for you, many software packages can make things easier for you. A spreadsheet software package only requires minimum skills in elementary mathematics and algebra. And if you'd like to be a programmer, math anxiety needn't stand in your way there either. It's just not that important.

Now that we've diminished the role of math anxiety in computerphobia, let's examine the other fears inherent in technophobia.

FEAR OF FAILURE

Why do so many of us suffer from fear of failure? Face it, ours is a success-oriented culture. Failure signifies self-defeat and disappointment for those who root for our success like parents, spouses, friends, bosses, co-workers, etc. Not too many individuals enjoy failure. But how badly can you "fail" with a computer? It's extremely difficult to break it (most computer errors are simple to fix; many computer breakdowns can be attributed to natural causes or mechanical failures). And while "failure" may be frustrating, it's hardly an awesome experience.

Dr. Gray of American University finds women's fears of computers ironic. "I think it comes from a general machine/electronic phobia," she says, likening technophobia to fears of car breakdowns. "That's sort of odd, because computers resemble typewriters more." And few women are afraid of typewriters. (Not liking them is another story.)

FEAR OF BEING REPLACED

Aha, now we've come to a legitimate fear. For it's a fact that robots can replace factory workers, computers can be programmed to replace middle managers, and new technology can make many traditional jobs obsolete. Yet hiding one's head in the sand won't save your job. If you're a secretary and your office is being automated, the event may be a lucky break for your career. If you embrace the new technology—study how it works and take on the responsibility of training co-workers—you're in line for a promotion to your first managerial or supervisory position. How does word processing center manager or trainer sound to you?

As illustrated in Chapter 7, technology opens new doors while it closes others. It's your chance to make the best of unavoidable changes in the marketplace. And continuing education is the key.

FEAR OF PRIVACY LOSS

A Louis Harris poll conducted in conjunction with a Smithsonian Institution symposium, "The Road After 1984: High Technology and Human Freedom," discovered that a majority of Americans view the computer as a threat to privacy and other individual rights. This *is* a valid fear, but one that shouldn't stand in the way of your career. If microcomputers enter your office, or the entire office is automated, what you put into your computer may become easily accessible to others. This just means you have to make an effort to protect yourself. If you're putting personal lunches on your company expense account, you're making a mistake, and you'd be stupid to include that information on your computerized calendar. As for someone taking credit for ideas you've typed into your computer, protect yourself with written memos to your superiors. Take credit for your ideas and for your work. (More on this in Chapter 7.)

FEAR OF COMPUTER DEPENDENCY

You've probably heard the term hacker used to describe a computer aficionado who sits glued to a terminal, night and day, addicted to the thrills of plugging into Defense Department information banks or corporations' financial data. Set your mind at ease. The typical hacker is young, male, and single-minded. Do you have an addictive personality? (Are you drawn to abuse of alcohol, drugs, or to another self-destructive habit?) Do you feel that your life is empty or that work should be all-consuming? If not, you hardly fit the hacker profile.

Computer dependency or addiction is possible, but chances are extremely slim. Yes, you can become dependent on a computer to the extent that if it breaks down, you'll be inconvenienced. But that's what happens when your car breaks down, or the plumbing gets clogged, or if your electricity, heating, or air conditioning system goes on the fritz. It's a fact of modern life. We rely on technology to make our lives easier, and if technology breaks down, we suffer a little. But only a little. Is it worth giving up your car, indoor plumbing, electricity, heat, or air conditioning? If you think so, then maybe computers aren't for you.

WHY BATTLE TECHNOPHOBIA?

Are you interested in any of the following?

- solving problems more easily
- saving time
- saving energy
- being creative
- understanding how to solve problems
- advancing your career
- making more money
- satisfying your curiosity
- translating computer jargon into English
- learning something new

If you answered yes to any of these questions, you're probably sufficiently motivated to conquer your technophobia—even if you don't realize it. Why? Because becoming computer literate or knowledgeable about technology can help you in all the above ways.

WHY IS MOTIVATION IMPORTANT?

Motivation is a critical term in psychology. It's also crucial to acting, and probably to succeeding at anything else as well. Motivation is what makes us tick. It might be a drive for success, achievement, or pleasure. Or it might be a strong force like guilt, fear, or curiosity. Think about your motivating forces. Why do you do the things you do? You probably have different motivations for different actions. Examine the list above under "Why Battle Technophobia?" Check off the reasons that are important to you. There's your motivation.

How do you capitalize on your motivation? If your motivation is to solve problems more quickly or more easily, then you will want to become computer literate. If you want to advance your career or make more money, you'll have to consider which positions appeal to you. (See Chapter 4 on what and where the jobs are, and their prerequisites.) If you just want to learn something new, how far you want to go in high-tech education will be related to the limits of your curiosity.

BATTLE PLAN

What's your next step? If you've already identified yourself as a sufferer, figured out what you're afraid of, and identified what motivates you to conquer your fear, you're ready to begin the journey to computer comfort. Recognize that:

- If you follow directions, you won't break the computer.
- Everyone feels nervous when trying something new.
- Your nervousness will pass.
- You can do it!
- Using computers will save you time and effort.

- Using computers can be fun!
- Learning how to use or program computers can open up exciting new career opportunities.
- High-tech skills can be your passport to a more flexible workstyle and lifestyle.
- High-tech skills can mean more money in the bank.
- Learning is an exciting, but occasionally frustrating, experience.

Enough pep talk. Before seeking training you'll need to decide what kind of high-tech career most interests you. If you're too constrained by technophobia to hazard a guess concerning your next move, a workshop on computer literacy should be your first step. If you're uncertain about what jobs exist and which career route would be the best for you, you would probably be best off starting with a seminar or brief course. By getting your feet wet in this way, you can discover if you'd prefer pursuing programming, systems analysis, database management, technical writing, engineering, computer design, teaching, or sales.

Talk to individuals who hold jobs in the areas that interest you. See Chapters 4 and 5 for more information on what and where the jobs are, and selecting the right job for you. Then get some hands-on experience using computers. See if you prefer using them as tools for common purposes, such as word processing, accounting or financial analysis, or teaching. Or would you be happier selling the equipment, repairing it, or even designing it.

Your preference alone won't guarantee you the right choice. Quizzes in Chapter 5 will help you match your likes, dislikes, aptitudes, and goals to a high-tech position. Before you're ready for that, read on for an overview of the kind of training available to you—to combat technophobia, learn how to use computers or learn basic programming techniques, particular languages, or other high-tech skills.

WORKSHOPS

Would a two-hour or one-day workshop on technophobia or computer literacy be sufficient to put an end to your fear? In a workshop you can expect to learn:

- computer buzzwords like "bytes" and "bugs"
- rudimentary information about how to operate a computer
- how to use software packages
- how a specific program works
- how a computer might save you time and money

A workshop can serve to break the ice or loosen the bars of your "cage" of fears.

If you're an independent sort, a workshop might be enough to ready you for continuing your education via self-instruction. However, if you're extremely phobic, you may want a little more hand-holding. And if your goal is not only to feel comfortable with computers but to program them, repair them, design them, or become a trainer, teacher, systems analyst, engineer, data base manager, or computer security specialist, you'll need further training.

Workshops are frequently available free of charge through computer stores, department stores, computer schools, public schools, private schools, trade and professional associations, and women's groups. Corporations may provide in-house workshops for their employees. Look for hands-on experience (actually using the computer), as well as helpful hand-outs (like a glossary of computer terms). Check the credentials of the workshop's leader. Can she or he speak English as well as "computerese"? Does the teacher know a great deal about computers, software, and their applications? How many people will be allowed to attend the workshop? (The smaller it is, the better.) Will questions be answered? What computer will be demonstrated? What software? Who is sponsoring the workshop and why?

If you can't locate a suitable workshop in your area, you might consider using CAI, computer-assisted instruction. If you're willing to interact with a computer on your own, this may be right for you, but you'll have to have access to a computer at home, at work, or at school. If you're not ready for CAI, you might try a workshop or seminar while attending a convention or while on vacation before proceeding to more independent book or hands-on learning.

SEMINARS

A seminar is more likely to offer individual attention as well as present more detailed information than a workshop. A seminar may last one day or less, or more. Ask what time will be devoted to hands-on experience, to answering attendees' questions, to a general introduction or overview? What are the leader's credentials?

If you're interested in a week or so of hands-on computer experience and individual training, you might consider attending computer camp. Computer camps are available for both adults and children. Computer camp instruction for adults is offered by Club Med (in such relaxing environments as Cancun, Mexico), Atari, and others. In this way, you could combine fun with education, sun with software.

MORE EXTENSIVE TRAINING

If you're interested in pursuing a technical career, a workshop or seminar would only be a preliminary step. Then you'd have to take at least a three-month training program. More likely you would need a two-year course or four-year degree. Before deciding you need to identify your career goals, the sources of financing (savings, loans, grants, scholarships) available to you, the time available to you, the degree to which you are flexible (can you attend school during the day? at night? on weekends?) and your geographical limitations (how far can you travel for education?). For a position as a computer service repairperson or technician, you'd require a minimum of three months training before continuing with on-the-job training. Generally, a two-year trade school certificate is preferred by employers.

If you're interested in programming, the minimum requirement would be a six-month certificate or comparable experience. In today's competitive job market for programmers, you'd be better served by a two- or four-year degree from a trade school, community college, four-year college or university. Most competitive would be a four-year degree from a prestigious computer science, information systems, or data processing department at a high-ranked university. Remember that the school, professors, equipment, courses, class size, and cost are all significant factors.

Look for professors who have experience in non-academic work. (Research or past business employment can provide them with a helpful network of contacts for job leads and internship possibilities for you, as well as a good sense of what the market demands.) Also look for equipment that's up-to-date and commonly used in the marketplace.

Similarly, a programmer should learn specific computer languages as well as the concepts behind them. Languages in heavy demand today are COBOL, FORTRAN, PASCAL, and C.

If your goal is to program in a business environment, be sure to take courses in management, accounting, finance, problem-solving, and entrepreneurship. A degree in information systems or business might be better for you than a computer science degree.

If you want to be a scientific programmer, a double major in computer science and a scientific discipline (physics, chemistry, biology, etc.) might be the best route for you.

An engineer would need a four-year degree for entry-level engineering positions and further education for more advanced research and theoretical work. Note that engineering as a course of study encompasses many disciplines: electrical, electronic, chemical, petroleum, ceramic, etc. Most in demand in the eighties are electrical engineers. However, best paid in 1983 were chemical engineers. Keep in mind that engineers face a severely economy-linked future. (See Chapter 4 for more details.)

OTHER TRAINING

Less technical high-tech positions include those of trainer, teacher, technical writer, computer graphics artist, and salesperson. How are they trained? Many of them receive all of their training on the job, but pre-professional training does exist.

Degrees in computer science, technical writing, computer graphics, engineering, merchandising, marketing, advertising, journalism, art, information systems, or data processing might provide the appropriate training for you. For more specific information on matching high-tech training and education to specific positions see Chapter 4.

What and Where the Jobs Are

Among the most commonly known high-tech jobs are those of keypuncher and computer and word processing operator. Yet these positions only scratch the surface of career opportunities available in the high-tech world. And since the majority of women employed in high-tech jobs are found in these lowest-paid of computer-related positions, the situation is ripe for a revolution.

More and more women—like yourself—are wisely attracted to the promise of high-tech careers. That's why it's critical to identify the position that offers you the appropriate blend of professional growth, financial compensation, and personal satisfaction, while best matching your interests, aptitudes, and training. Chapter 5 will help you discover which of the many positions explored in this chapter provides the optimum fit for you. Chapter 6 will assist you in learning the nuts and bolts—or bits and bytes—of beginning your high-tech career. And Chapter 7 will guide you upward in your chosen profession, while Chapter 8 offers a primer on starting your own high-tech business. But first, in this chapter, you'll want to familiarize yourself with the smorgasbord of career opportunities on high-tech's groaning board.

WHO ARE THE HIGH-TECH EMPLOYERS?

Who employs high-tech workers? Computer and computer-related manufacturers (makers of mainframes, micro- and minicomputers, peripherals and parts for business, personal, and educational use) and software houses are two prime employers. So are more specialized companies involved in the smaller fields of robotics, fiber optics, lasers, genetic engineering or biotechnology, CAD/CAM (computer-aided design/computer-aided manufacturing), high-tech medical devices and testing equipment, ergonomics (the science of designing products for safe, efficient, and comfortable use by humans), and artificial intelligence. In addition, business is booming for other high-tech employers in the fields of telecommunications, office automation, computer repair and maintenance, and computer graphics.

Education accounts for yet more job openings: in elementary and secondary schools, trade and vocational schools, colleges and universities, adult and continuing education programs, private and in-house training centers, computer camps, CAI (computer-aided instruction) programs and other educational software producers. Add to this all the computer-related and data or information processing positions in such traditional industries as:

- banking
- insurance
- real estate
- financial services (accounting, investment analysis, venture capitalization)
- medical services
- legal services
- other services (data processing, timesharing, management consulting)
- manufacturing (electronic goods, aircraft, and low-tech items produced by high-tech methods or controlled and tracked with them)
- transportation
- utilities
- publishing and communications
- entertainment and the arts

- building design and construction
- retailing
- the wholesale garment industry
- government (The federal government is the largest user of computers, particularly in the Department of Defense and the National Aeronautics and Space Administration. State and local governments use computers for libraries, public schools, and administrative offices.)

In many cases you have a choice of position *and* industry, options in responsibilities and working environment. You could be a programmer at a large bank, a small software house, or in a medium-sized school district. You could use computer-aided design tools in constructing office buildings or designing auto parts. Location will affect the opportunities available as certain industries are more prevalent in particular areas of the country. And economic conditions and the degree of labor competition vary from locale to locale. (A higher unemployment rate and a greater number of high-tech educational or training programs in an area will heighten the competition you'll face for jobs.)

WHERE THE JOBS ARE

Before examining specific positions and hearing what women who hold these jobs have to say about them, let's look at the larger picture. In the United States, Silicon Valley draws most of the publicity as the country's high-tech capital. Silicon Valley is the name given to a twenty-five-mile area in California's Santa Clara Valley, which stretches from Palo Alto to San Jose. It's home to more than 1,400 electronics firms, and employs engineers and computer designers as well as technicians, assemblers (factory line workers), mechanics, and office workers.

A woman needn't go west, however, to secure a high-tech job. Opportunities exist throughout the United States (and in other countries as well), with more opportunities available in urban areas, particularly in the larger cities. Outside of Silicon Valley, the following areas are considered the hot spots for high-tech employment: New York metropolitan area (New York City; New Jersey; Westchester County, New York; Fairfield County, Connecticut); greater Boston area (along Route 128 and spreading out to New Hampshire and Vermont); Washington, D.C., metropolitan area, Gaithersburg, Maryland; Houston, Dallas, and the area stretching between Austin and San Antonio; Phoenix and Tucson; and Philadelphia.

POSITION VS. INDUSTRY

Which should you select first, a desired position or a preferred industry? It's not an easy question to answer. The industry, more than any specific company, can set the tone for your employment or working environment. The old-guard banking and insurance industries demand professional interaction with others according to rather formal and conservative rules of behavior. This will affect how you'll have to dress and communicate to succeed. On the other hand, the maverick atmosphere of a young, ambitious software house will be more accommodating of individual differences, more casual in style and structure (less hierarchical), but more than likely also more demanding of your time, energy, and enthusiasm.

Take into account how you could be affected by an industry's instability (shakeouts in personal computer manufacturers are just one example, the recession in the automobile industry is another). And keep in mind that an industry's growth or lack thereof does not necessarily predict a company's nor an individual's.

THE NAME GAME

Often the same positions bear different names in different industries and even at different companies. For example, someone who repairs or maintains computers may be called a computer mechanic, a computer technician, or a field engineer, depending on the employer and the particular client served. All too frequently, openings labelled with the same occupational title will themselves demonstrate many distinctions upon close examination. A programmer is not always a programmer. There are applications programmers and systems programmers. There are mini/micro programmers and mainframe programmers. There are programmers who specialize in FORTRAN (a computer language), and others versed in COBOL. There are programmers who maintain existing programs and "debug" new programs and programmers who "write code." And another class of programmers supervises other programmers and coordinates projects. Programmers may be called programmer trainee, or junior, senior, lead, or chief programmers to indicate their level of experience; or project leaders or managers depending on their managerial responsibilities.

The best way to compare high-tech positions which bear the same title is to compare the job descriptions. What exactly are the responsibilities of the job holder? The prerequisites?

You'll find in the following pages detailed job descriptions for the most promising high-tech jobs available today. And you'll read first-hand accounts of what these jobs are really like from women presently employed in these positions.

Computer Service Technicians

High-tech careers are not closed to individuals who have only high school educations. Those who are uninterested in pursuing further formal education might pursue careers as computer service technicians or computer mechanics. (Other titles include customer engineers and field engineers.) Frequently, however, one or two years of electronics or engineering training past high school is required by employers.

PREREQUISITES AND RESPONSIBILITIES These positions require technical ability and an analytical mind. Computer sytems must be serviced or maintained periodically (adjusted, cleaned, lubricated, and tested), as well as diagnosed and repaired when a problem occurs.

When a computer is "down," or not functioning properly, work can come to a halt. Down time can mean money lost, deadlines missed, project overruns, and a lot of hair tearing and cursing. The computer mechanic plays a vital role in company operations and therefore may experience a substantial amount of pressure.

A computer mechanic must be able to:

- operate computer and related equipment
- understand technical manuals
- diagnose malfunctions
- use tools like voltmeters, ohmmeters, oscilloscopes, soldering irons, and other hand tools
- repair breakdowns
- order parts and keep inventories of parts
- listen to customers' complaints and answer questions
- keep up-to-date concerning maintenance and repair techniques
- occasionally train and supervise other technicians

THE HARD FACTS A computer mechanic requires good eyesight, patience, and a logical mind. The market for computer mechanics is strong. There were 83,000 computer technicians in 1980, most of whom were employed by service companies and manufacturers, and that number is expected to grow. The average starting salary is $14,000 a year, increasing to $20,000 to $29,000 for experienced workers. This position is overwhelmingly dominated by men, but results count more than sex.

Programmers

To put it most simply, programmers tell computers what to do by providing them with a specific, precise, and logical series of step-by-step instructions. The end-product—what programmers produce—is called a program or *software* (package). The program allows *end-users* (the individuals who actually use the program) to accomplish tasks easily and efficiently using the computer (hardware) in conjunction with the software. With the right software, users don't need to be programmers themselves to do word processing, financial analysis or management, data filing and management, computer-aided instruction, or medical and psychological diagnosis. The efforts of programmers vastly simplify the procedures followed by computer users. Complex tasks that formerly required days, weeks, or more can now be accomplished within minutes or hours.

WHAT PROGRAMMERS DO Computers, unlike most people, need to be told *exactly* what to do in order to perform properly. And this information must be provided in the computer's own language—or in a language the computer has been programmed to "understand" (that is until artificial intelligence makes greater headway). A programmer, no matter what her specialization, has four basic functions (although these functions may be divided among different individuals at certain companies):

1. to understand completely the task that a computer is being called upon to perform (such as establishing an automated quality control system for a computer manufacturer) by consulting with systems analysts and end-users or clients to determine their needs and preferences
2. to prepare a *flow chart,* a diagram of all the steps to be followed to accomplish the task, with each step following logically from the previous one
3. to code the information in the appropriate computer language as efficiently as possible (Computers don't understand English. They can understand languages like COBOL, FORTRAN, or BASIC, so each step in the flow chart must be translated into—or coded in—a few "words" of the computer language. A com-

plex task can require many, many lines of code to program. Since these lines take up "memory," the fewer lines the more efficient the program.)

4. to test and debug the program (Typically, a program does not work properly the first time around. There will be errors in the coding, either by the use of an incorrect word or symbol or an error in logic, like skipping a step. Locating and fixing the errors —*debugging*—can require as much time as writing the program, if not more time. Often programmers also prepare or help prepare instruction sheets for the computer operator or end-user. (This may be done with a technical writer or documentation specialist.)

A junior-level programmer or programmer trainee might concentrate on writing code or maintaining old programs. A programmer fresh from school is "totally useless" until she has a year of on-the-job training, according to Diane Hajicek, chief programmer at MicroPro International, Inc., a California software house. These programmers will be closely supervised. More experienced programmers will have responsibility for the full range of programming functions. Senior, lead, or chief programmers may work completely independently, or with other programmers in group projects or in consultation with systems analysts, consultants, and end-users. The senior programmer helps direct and manage programmers assigned to specific projects. She will be involved in staff selection, scheduling, budgeting, coordinating, and evaluating work. The lead programmer has more extensive administrative responsibilities in organizing projects, assigning staff, preparing schedules and budgets, assessing progress and problems, developing and enforcing programming standards, and recruiting, interviewing, hiring, reviewing, and counseling programmers. In some companies, senior or chief programmers do not have managerial responsibilities. Their titles indicate seniority and experience.

COMPUTER LANGUAGES The computer language or languages you learn will bear directly on the type of programmer you can be and the industry in which you'll be employed:

- COBOL (Common Business-Oriented Language) developed by Grace Hopper (see Chapter 1 for more information), is the language used most heavily in business applications. It's considered a high-level language.
- FORTRAN (FORmula TRANslation), also in common use, has mostly scientific, technical, and statistical applications. It's also a high-level language.
- BASIC (Beginner's All-purpose Symbolic Instruction Code), which may be the easiest language to learn—and the most unsophisticated—is the language of many microcomputers.
- BAL (Basic Assembly Language) and ALC (Assembly Language Code) are assembly or low-level languages, languages that control the internal operation of a computer. These are used by systems programmers.
- ADA, a relatively new language, was commissioned by the U.S. Department of Defense to replace FORTRAN, and named after Lady Ada Augusta, the first "programmer." (See Chapter 1 for more information.) ADA involves using "objects"—prewritten software—in combination to assemble a complex program without writing original code.
- PASCAL, often used for business programs, was named after a seventeenth century French mathematician and philosopher by Niklaus Wirth, a Swiss computer scientist. It places more restrictions on a programmer, but because it's very orderly, complex programs can be written relatively quickly.

PROGRAMMING SPECIALTIES Programmers are either systems programmers or applications programmers. Systems programmers must have a detailed understanding of the inner workings of the computer, along with knowledge of low-level or assembly languages. Systems programming is more technical than applications programming (and more male-dominated as well). Using languages like BAL or ALC, a systems programmer instructs a computer how to understand high-level languages like COBOL or FORTRAN, or how to operate peripheral equipment like printers and modems. Systems programmers tend to earn higher salaries than applications programmers, but their hours are often less regular. (Systems programmers in data processing departments may have to solve problems with systems software when the computer is not being used for applications—that is on weekends and holidays.) Systems programmers at computer manufacturing companies would not face this problem. An applications programmer specializing in scientific or technical applications might move into systems programming.

Other possible programming specializations include:

- mini- and microcomputer programming
- distributed data processing systems programming (where small computers or terminals in different locations are "connected" so that they can communicate with each other)
- programming for time-sharing computer systems (where a large computer is shared by end-users at different companies for varied purposes)
- video game programming
- programming robots
- educational programming
- programming to help the handicapped
- defense and weapons programming
- specific business applications

While it's true that many employers consider the knowledge of a specific language of less import than an understanding of the concepts behind it (which enables a programmer to pick up other languages easily), specializing in a computer language—or learning one with limited applications—can have a great effect on your career direction. If you know a language in demand, you'll find more employment opportunities. And if the language you're studying is edging toward obsolescence (as some argue is the case with BASIC), you'll be less desirable in the marketplace.

Certain languages will lead you to employment in specific industries and geographical areas. For example, assembly languages are in demand for systems programmers at computer manufacturing companies. And these companies are found largely in California and Massachusetts. The demand for ADA, critical to U.S. government defense work, is dependent on the state of the U.S. military budget, and this work is localized in the Washington, D.C., metropolitan area and in California.

If you specialize in COBOL, you could be a programmer in a business environment (bank, insurance company, large corporation) working in such areas as payroll, data base management, information processing, recordkeeping, billing, inventory control, accounting, or financial analysis.

A scientific programmer specializing in FORTRAN or APL (A Programming Language) might create programs that:

- do numerical simulations for use by astrophysicists studying sources of radiation outside the solar system
- do velocity analysis for geophysicists searching for petroleum
- provide graphics representations for biological conditions, such as in medical imaging equipment
- determine flight paths for air traffic controllers

A mathematics or science background is helpful, and often required, for scientific programmers.

You could be employed by a small, medium, or large-sized firm or by the government in a data processing or *MIS* (management of information systems) department. Or at a software, hardware, consulting, or other computer-related service company. (At a small company you would more likely be hired as a programmer analyst; at a large company, as programmer trainee.)

You could do *custom programming*—creating or adapting software to a client's or user's particular needs—or simply maintain programs. At a consulting or service company or at a software house, a programmer would be a more valued employee because she would contribute more directly to the company's bottom line. At these companies, programmers hold *line* positions, not *staff* positions (for more on this, see Chapter 7).

JANET OWENS Janet Owens worked as a programmer for an oil well company in Houston for two summers while she earned a B.S. in computer science. After she graduated from college, she went to work full-time for the same company. Her department included fifteen engineers and five programmers. She learned on the job about drilling engineering and participated in a six-week course the company arranged for new programmers. When she moved to New York with a year's programming experience, Owens landed a job at a software house designing a software package to do recordkeeping and portfolio evaluation for stockholders and investors. She had no previous background in financial analysis before her present job. And in Owens's present position there is no formal training program. She learns by asking questions.

What traits does a programmer need for success? Owens names the following: ability to simplify things, logic ability, patience, desire to teach, and understanding that the user is not a "computer person" and that the program must be designed with user's needs and preferences in mind.

Owens moved from her first job to a second in a completely different industry and geographic area. And while she studied FORTRAN in college, she now programs in BASIC. "The logic is similar, although the syntax is different," she says. This flexibility is both a prerequisite and a bonus for programmers. Programming experience is a valuable commodity that can be parlayed into various career directions.

While Owens notes her job is stressful, she appreciates the flexibility—not uncommon for programmers—in arranging her own hours (although they average ten per day) and working "totally independently. . . . I wouldn't have it any other way," she says. Does she foresee staying with programming? After two years of programming, she is interested in CAD/CAM.

DIANE HAJICEK Diane Hajicek is a chief programmer at

MicroPro International, Inc. in San Rafael, California. She created Datastar, a general purpose data entry program that controls information so that it's accurate and can be entered into the computer quickly. Hajicek has been at MicroPro since 1980 and was the tenth person hired at the company. Starting at an up-and-coming software house allowed for greater flexibility in negotiating royalties or equity. And these are the two arrangements by which a programmer—free-lance or staff—can directly benefit from a product and company's success and profits.

Hajicek has a computer terminal at her home and often works there. She, like Owens, sets her own hours and very rarely works with others. Hajicek is not interested in pursuing the management route. (She's supervised others in the past and doesn't believe she has "people skills.") Instead, she sees herself on a technical career path, remaining in programming, where she believes she can be more independent and make more money.

Hajicek appreciates the "maverick" atmosphere at software houses. She observes that programmers who work with IBM computers and COBOL in industries like insurance and banking tend to hold more structured, nine-to-five positions. Her diverse background includes stints at an insurance company (while an undergraduate, she majored in math and minored in computer science), graduate school for chemistry (she left when she decided chemistry wasn't for her), Bell Computer Products (as a junior programmer working on languages and operating systems), a bank (her work was contracted through a "rent-a-programmer" company; she worked on a data entry system); and Imsai (manufacturer of the first microcomputer, sold in kits, where she progressed from systems programmer to chief programmer).

Hajicek's career, like Owens's, illustrates that many different programming career paths exist. You needn't travel in a straight line, remaining in the same industry or continuing at the same type of programming. You can shift from industry to industry and from systems programming to applications programming. A programmer can also make the transition from programming to systems analysis, training, sales, technical writing, marketing, data base management, computer security, and managerial positions.

Hajicek has not taken additional courses since she left school. She's been able to learn all she needs on the job. The best way to learn programming, she believes, is to "read other people's programs and learn their techniques."

DEMAND Programming may be the right position for you. (The next chapter will help you test your aptitude and interest.) It can be an entry-level or advanced position, on a staff, free-lance, or consultant basis. In 1970 there were approximately 161,000 computer programmers in the U.S. By 1980 the figure had more than doubled to 341,000. The Bureau of Labor Statistics predicts a sizeable 47 percent increase to 500,000 by 1990. One of those programmers could be you.

SALARY You can expect a starting salary between $13,000 and $20,000, slightly higher for scientific and systems programmers. And salaries can increase rapidly for programmers who learn quickly. After five months at her first job, Janet Owens received a 15 percent raise and then a 10 percent raise after six months at her second job. (Employers will pay to hold on to valuable high-tech employees when the market is tight.) According to Source EDP, a high-tech recruiter and placement service, the median salary in 1983 for commercial programmers with more than four years experience was $31,300. For engineering and scientific programmers with equivalent experience, the figure was $33,800.

You should be aware that some high-tech companies are moving toward the use of automatic systems generation (ASG). This enables a programmer to "customize" a system without having to write a program from scratch. It turns a programmer's job into more routine, assembly-line type work. Companies that use ASG claim greater productivity—and pay programmers less. Also, companies that rely on prepackaged software have less need of a programmer's technical skills. Some labor experts predict a declining need for programmers due to these changes. Now for the good news: the demand is expected to be strong well into the next decade.

Programmer Analysts

The programmer analyst position, which combines functions of the programmer with that of the systems analyst, can be an intermediate position or synonymous with the systems analyst position. Some high-tech companies hire programmer analysts; others clearly separate the positions. Rather than offer a separate job description here, I leave it to the reader to combine the descriptions of programmers' and systems analysts' responsibilities.

Systems Analysts

Systems analysts—sometimes called systems engineers—are basically problem solvers. Like programmers, they specialize in either scientific or commercial applications. No matter what the problem at hand, the systems analyst asks questions like:

- How is the job being performed now?
- How much time does this take?
- What are the costs in labor, salaries, equipment, maintenance?

WHAT A SYSTEMS ANALYST DOES It's the systems analyst's job to discuss and analyze a problem with managers or specialists to pinpoint what needs to be done to develop a new system (what information must be collected and what form it should take, what equipment is needed and what tasks the computer should perform); such techniques as cost accounting, sampling, and mathematical model building are used. A systems analyst also translates the system or solution into charts and diagrams to describe how it will work to managers and customers and sometimes prepares a cost-benefit analysis to help a client or department determine the suitability of the system proposed. Systems analysts may translate the requirements of the system into capabilities that the computer hardware can handle and prepare specifications for programmers to follow in instructing the computer how to process information, and work with programmers on testing and debugging.

Systems analysis work is complex and varied. A systems analyst might work on improving a system already in use by simplifying procedures, adding or combining functions to be performed, adapting the system for use with other machinery, or doing more theoretical research—as in advanced systems design—to develop new methods of systems analysis. A systems analyst's goals are always to increase efficiency and productivity.

PREREQUISITES Systems analysts may have backgrounds as programmers or, in the case of commercial systems analysts, in business management, accounting, or economics. Scientific systems analysts often have backgrounds in mathematics, engineering, physical sciences, or computer science, information systems, or data processing. Prerequisites generally include:

- familiarity with programming languages
- understanding of computer concepts and data management
- prior work experience (almost 50 percent of the individuals who become systems analysts start their careers in other occupations, typically programming)
- ability to think abstractly and conceptually
- ability to interact well with technical staff, those without computer backgrounds, all levels of management, and clients or users
- detail orientation
- problem-solving and analytical ability
- communication skills, oral and written
- ability to work on two or more projects simultaneously
- organizational and management skills
- teaching and persuasion abilities

Programmers may become systems analysts, but so may industrial engineers and management analysts. The major distinction between programmers and systems analysts is that systems analysts must concentrate on the user's needs and how to meet them; programmers on writing software. A systems analyst must be able to understand a problem thoroughly—whether it be at a manufacturing company, service organization, government agency, bank, insurance company, retail or wholesale business, hospital, or school—and be able to recommend an appropriate cost-effective solution. This involves:

- determining what ultimately the user wants to accomplish and comparing it to the process presently in use in terms of what and how equipment, people, time, and other resources are used; understanding the scope of the problem
- developing alternative methods of accomplishing the task, comparing them and selecting the best one; perhaps performing a feasibility study
- recommending a system, explaining it, trying to sell it by promoting its best features (such as increased productivity, long-term financial savings, compatibility with presently owned equipment)
- working out specific details for implementation of a system if approved

GROWTH OPPORTUNITIES While the systems analyst's job is highly technical, it is also advantageously positioned for learning a company's internal operation at its highest levels. It can be an excellent stepping stone for movement into high-level general business management positions for the systems analyst interested in entering management. (Financial rewards will be greater, but management is not for everyone.)

Systems analysts more satisfied with the technical side of their positions can move into senior and lead systems analyst positions. They will have to continue their technical education to keep abreast of new concepts, advances in equipment, new methods of information processing, and new developments in the particular industry in which they work.

In structuring a career path, the systems analyst must decide whether to follow a technical or managerial route. You can progress only so far as a systems analyst if you're unwilling to take on managerial responsibilities. (An in-depth discussion of this issue can be found in Chapter 7.)

TRAINING Systems analysts are generally college graduates with training in the particular industry they serve. An M.B.A. can be a very helpful degree for further advancement. Systems analysts may work in computer, data processing, or MIS departments, or in consulting or service companies. Clients may be small, medium, or large-sized firms automating tasks for the first time or "enhancing" equipment or processes with the assistance of the latest technological developments. (In fact, even when the systems analyst is not, technically, a consultant, she functions like one. More on the consultant's role later.) Knowledge of business or science, hardware, software, compatibility, and systems programming all come into play. Yet technical knowledge and skills are not sufficient by themselves for the successful systems analyst. People skills are critical to the systems analyst's role as:

- interviewer
- researcher
- teacher
- salesperson
- leader/supervisor/manager
- team member
- listener and communicator

TECHNICAL VS. MANAGERIAL ANALYSIS An average breakdown of the systems analyst's job is 60 percent technical, 30 percent managerial, and 10 percent translation to written and illustrated forms. In some companies the systems analyst position is primarily technical. In others, the holder of this position is mostly a salesperson. At computer manufacturing companies, the systems analyst's functions may be performed by a computer designer or systems designer or computer architect.

The systems analyst position might appeal to you if you are

dissatisfied with the more limited role of a programmer, desire to be involved in a project from start to finish, want to interact with others at all levels of a company, want to be innovative and change systems rather than follow instructions, want to be a generalist more than a specialist, want to combine technical and managerial skills, like working well under pressure, or are interested in advancing to upper management. Travel is often required for meetings with clients and vendors, and for attendance at trade shows and professional educational workshops or seminars.

The systems analyst job itself is not new to the high-tech world. The position's tools and methods are what's new. Before computers were ever used to automate business and scientific processes, systems analysts (at one time called efficiency experts) worked on improving operations for greater efficiency and productivity—manually. High-tech developments in hardware and software have shifted the emphasis in systems analysis to the use of computers in furthering those same goals.

Systems analysts hold high-pressure positions. There is a great deal of money riding on their recommendations. The wrong system—one that fails to perform properly—represents a very costly error.

THE HARD FACTS Good systems analysts are in great demand. In 1980 there were 205,000 systems analysts in the United States. The BLS predicts a 95 percent increase by 1990 to 400,000. The average starting salary is $17,000 to $22,000. In 1983 senior systems analysts earned a median salary of $34,700 after four years, according to Source EDP.

Technical Writers

Technical writing is a fast-growing career that's a custom fit for the individual with writing talent, an interest in technology, and a bent toward education. The technical writer's primary task is to translate scientific and technical information into language easily understood by a lay audience. This can involve researching, writing, and editing to produce users' manuals, promotion materials, public relations material, instruction sheets, fact sheets, marketing brochures, advertising copy, company newsletters and publications, technical columns and features in trade and popular publications and newspapers, technical books, and audiovisual materials.

It's the technical writer's job to explain, describe, document, and promote all the latest high-tech products. Therefore, she must be able to understand enough of their technical aspects to explain how to use them, describe their features, and promote their sale.

PREREQUISITES Some organizations only hire technical writers who have technical backgrounds. Others firmly believe that writing ability and an understanding of the user's point of view far outweigh the significance of a technical background. These employers, like Rolf A. Fuessler, vice-president of corporation communications at Camp Dresser McKee Inc., an environmental-engineering company, believe "it is far easier to turn a writer into a technician, than the other way around." At Wang Laboratories near Boston, the word processor and computer manufacturer employs about ninety writers and twenty-five editors who write and edit instruction manuals and training books and translate them into foreign languages. An employee claims that not one has a degree in computer science.

As readers of early user's manuals can attest, much of the instruction material initially written for personal computer and software buyers was written by technical employees with little understanding of "non-computer" people. Poorly written, badly organized, and unclear manuals, advertising, and marketing material spelled doom for many companies in the marketplace. Smart manufacturers realized they needed to *explain* their products to potential customers clearly—and to *sell* them. (Early high-tech manufacturers were used to dealing with hobbyists or data processing personnel, "techies," who needed less explanation and less marketing attention.) This signalled the rise in demand for technical writers and marketeers.

THE NAME GAME Technical writers may be called documentation specialists, staff writers, publications engineers, communications specialists, industrial writers, or instructional materials developers. They may specialize in writing manuals for computer operators or users, proposals, marketing, advertising, or promotion materials, annual reports, company newsletters, speeches, material for displays and exhibits, press releases, or catalogues. Or they may cover the industry for trade magazines, newspapers, industry newsletters, consulting companies, brokerage houses, or in books.

WHAT A TECHNICAL WRITER DOES Technical writers are concerned with describing, explaining, educating, and promoting. They must understand their reading audience thoroughly and write in a style appropriate to that audience. They may write to business users, home users (families, children, singles), technically aware people, unsophisticated users, high-level management, low-level staff (operators), all levels of an organization, government agencies, venture capitalists, bank officers, potential investors, and stockholders.

Generally, technical writers function as part of a team. They may work closely with programmers, systems analysts, engineers, technicians, accountants, illustrators, printers, production staff, editors, photographers, salespeople, and managers. They may be responsible for coordinating all efforts to produce a final product, whether a manual or an ad, and meet a deadline:

- researching and assembling or organizing necessary information—studying mock-ups (if the product doesn't exist yet) or product samples, reports, drawings, charts, specifications, descriptions; consulting with technical staff
- outlining information
- preparing a rough draft
- working with a designer on layout, use of charts, tables, illustrations, photographs, captions, titles, headlines, print style

- revising draft until satisfactory—consulting with technical staff, managers, editorial team members
- coordinating production
- editing and proofreading material

EDITING VS. WRITING Some technical writers do more editing and rewriting than actual writing; others are allowed greater creativity and flexibility. The more technical the material and the audience, the more rigid the writer's guidelines. Advertising and marketing technical writers tend to be allowed the greatest flexibility. For all technical writers, however, providing information or presenting a message clearly takes precedence over rhetorical flourishes. The poet or prose stylist should think twice about pursuing a technical writing career. Ask yourself if you can adapt your writing style to the job's requirements.

EMPLOYERS OF TECHNICAL WRITERS Technical writers may be staff members or employed by:

- manufacturers of computer hardware, software, computer peripherals and parts, aircraft, weapons, medical diagnostic equipment, genetically engineered products (like insulin, seeds)
- trade and professional associations
- public relations companies
- advertising agencies
- public and private research and development laboratories at educational facilities, hospitals, pharmaceutical companies, electronics firms, energy development and conversion organizations, environmental health and safety concerns
- publishing houses
- consulting and service companies
- government (75 percent of the U.S. government's 1,700 technical writers work for the Department of Defense on manuals regarding the construction, maintenance, and use of weapons and instruments; NASA also employs technical writers)
- investment houses and newsletters

- trade magazines and newsletters
- popular magazines
- newspapers

TRAINING Educational background can run the gamut from English or journalism to computer science, from history to philosophy. What counts is clear writing ability and a grasp of technical information. Degree programs in technical writing or communications are recent developments on the educational scene. In 1980 there were only ten colleges and universities with such undergraduate programs. Most technical writers begin their careers as technicians, editorial assistants, writer trainees, junior writers, or advertising, marketing, or public relations assistants.

Technical writers need the following:

- curiosity
- dedication to precision and accuracy
- a logical bent
- independence (ability to work unsupervised)
- self-discipline
- patience and persistence
- tact and cooperativeness
- appreciation of the power of words
- ability to put themselves in the "shoes" of their readers
- organizational ability

ADVANCEMENT The technical writer rarely has a great many managerial functions. She may progress to running the editorial, publications, or communications division of a high-tech company —or the marketing, advertising, or public relations department. Or a technical writer could move into a customer support or training position. A lucrative way of making these moves would be to become a consultant or a free-lancer, leading writing improvement seminars for industry and government employees, or writing on contract—especially in syndicated newspaper columns or popular magazines. A technical writer might also move into sales, using oral communication skills more than written ones.

DEMAND The demand for technical writers is predicted to rise because of the increasing importance of marketing to the success of high-tech companies, the explosion of technical information and products, and the growing complexity of technical developments.

"There are lots of career opportunities in computer journalism," observes Maggie Canon, editor-in-chief of *InfoWorld.* "There seems to be a new publication every day, and they're looking for people who have some knowledge of computers and can write. And there is no prejudice against women."

SALARIES Technical writers started at an average salary of $15,200 in 1980, according to a study undertaken by the American Management Association. Experienced technical writers earned between $17,000 and $25,000. (Editors earned more, from $21,000 to $31,000.) Salaries depend on education, experience, responsibilities; type, size, and location of employer.

Sales

One of the more lucrative high-tech career routes is through sales, whether the position is as a manufacturer's sales representative, a computer store salesperson, a software outlet salesperson, or a computer services company (time-sharing, consulting) marketing representative. Salaries are frequently based largely on commissions and the proficient salesperson can do very well. In high-tech sales, you could sell such diverse products as:

- mini and microcomputers
- mainframes
- business software
- video game software
- educational software
- computer peripherals—printers, modems

- computer accessories—disks, paper, cartridges or ribbons, printed forms, paper, punch cards, magnetic tape
- telecommunications equipment and services—PBXs, videotext, telephones
- high-tech medical equipment like CATSCAN (computerized axial tomography)
- genetically engineered products like Genentech's insulin
- lasers
- ergonomically designed furniture—desks, chairs, tables, file cabinets, lamps
- computer books (technical and popular)
- computer insurance
- computer training (at vocational schools salespeople may be called "counselors" but they earn commissions on their sales)
- time-sharing services
- analysis of company management, industry, investments, finances
- computer repair, maintenance, and service
- information on data bases
- used computers
- advertising space for high-tech magazines and newspapers
- consulting—design, programming, systems analysis, etc.
- employee recruitment services
- trade shows
- job fairs
- training programs—seminars, CAI programs, workshops, audiovisual programs
- word processing services

The list is long, and this statement, an understatement.

WHAT A SALESPERSON NEEDS Whatever the product, as a salesperson you must be knowledgeable about it—and about its competition. The successful high-tech salesperson usually doesn't need a very technical background. Training and technical support is often supplied by other individuals. More important are:

- articulateness
- good communication skills (listening, speaking, writing)
- powers of persuasion
- enthusiasm
- motivation
- resiliency
- good self-presentation skills
- confidence
- assertiveness
- likability
- organizational abilities
- the ability to analyze a customer's needs and how to meet them
- initiative
- independence
- quick learning ability

BACKGROUNDS Many high-tech salespeople begin their careers in other positions. It's not uncommon for trainers and customer support people to move into sales. In addition, many high-tech salespeople have switched from low-tech products—"trading up"—for greater financial rewards. Many women, too, have correctly targeted sales positions in high-tech fields as promising, and made the transition from office work and teaching.

CHERYL ASHE After more than three years as a secretary, Cheryl Ashe realized that she faced a dead end in her career. While she had taken the initiative to learn how to operate the computer graphics system in the office, her job never progressed much past training engineers how to use the system and entering in data. She decided there was no place to go and took a new job that combined secretarial functions with those of a computer operator. Eventually she moved into programming, but she felt the pay she received was still insufficient. Her next move: sales.

Ashe started her own business, selling art for new buildings. It involved a lot of cold calling, a lot of work, but it was very lucrative. Nevertheless, after a year, Ashe was ready to move on. Why? "I didn't want to work for myself anymore," she says, "because I didn't have anybody to talk to."

Ashe made the rounds of employment agencies, telling them exactly what she was looking for: a position in computer sales. "I had worked on all the large word processing systems, and I could relate to the people I'd be selling to." She landed a job as account executive at a computer store. After a year she says, "I definitely feel this is the marketplace to be in."

She has a monthly quota to meet and is paid on a commission basis only. Ashe believes this allows her to make more money than a base salary would. "You have to be with a company that's carrying the right kind of machines and is service-oriented," she cautions.

Also important, says Ashe, is building a client base, increasing sales by satisfying customers and receiving referrals. She thinks that after the first six months needed to learn about the equipment and get established, a computer salesperson can expect to earn between $30,000 and $40,000 annually. Career paths include promotions to sales manager, assistant manager, and store manager.

SALES RECORDS Frequently high-tech employers demand evidence of sales ability such as experience and a proven sales record with like products. However, with the list of new, innovative high-tech products ever increasing, exact experience is hard to find. Employers will often accept any sales experience or make hiring decisions based on their assessment of your sales "personality."

Typically, companies hand down quotas for their sales force to meet—with rewards (bonuses, trips, awards, promotion to management) for surpassing them, and penalties for failure to meet them (warnings and eventually ouster). The atmosphere may be competitive or supportive. Training may be stressed or stinted. Leads may be provided or all calls may be "cold." The bottom line for salespeople is their sales record.

As Maggie Canon, editor-in-chief of *InfoWorld,* remembers, if you can't meet your quotas, you're out. On her zigzag path to her present position Canon sold printing services before moving to microcomputer and dedicated word processor sales. Unfortunately, her success with printing services didn't carry over to high-tech products. ACS, the microcomputer manufacturer, went out of business after six months. And at the word processor manufacturer, Canon was not meeting her quota and realized that sales wasn't "up her alley." She reevaluated her career direction and decided to build on the knowledge she had picked up about how a computer was built, programmed, and used. Her next stop: selling herself for her first job in computer journalism.

Sales success doesn't always carry over from one product to another, and from one company to another. Yet risk-taking is often the name of the game for a successful salesperson. If you're interested in sales, it's critical that you thoroughly research the company you'll be representing and the product you'll be selling. Your success depends on the company's—and the product's—competitiveness and strength. If it's a new product, does it fill a niche in the marketplace? Is it sufficiently competitive with other products or services on the market? If the company is well-established, what is its reputation in the field? Would you be comfortable selling to its customers, be they small businesses, large corporations, individuals, schools, hospitals, government agencies, bookstores, law firms, accounting concerns, travel agencies, food service operations, etc.? Will you be paid a regular salary, on a commission basis, or by a combination plan? Which would work best for you? And last, would you be content to represent the particular company?

The high-tech sales position has gained in importance as high-tech management has realized that having a good product or providing a good service is no guarantee of success. "Because we are maturing as an industry, it's becoming more marketing driven, rather than technology driven. And I think women are moving into the business area of marketing more easily than they are into the engineering aspect of it," Maggie Canon adds. This shift promises a rosier than ever future for the talented high-tech salesperson.

THE HARD FACTS The exact number of high-tech salespeople is difficult to pin down because the total pool of salespeople encompasses so many different industries and products. Nevertheless, the number is expected to increase as new products and service companies demand sellers to connect with buyers. Salaries vary tremendously, depending on the item sold and an individual's sales record. The higher the ticket price, the more lucrative the sales position. Many successful high-tech salespeople choose to stay in sales rather than move into management where training, supervising, and other managerial skills are called for. The salesperson who lacks interest in management can advance by tackling a larger territory, additional or other products or services, or moving into consulting.

Customer Support Representatives

Customer support representatives may be called tech reps, sales engineers, systems engineers, or marketing technical support representatives. As the titles imply, this position is primarily a technical one, although training may be primarily on the job. Like the high-tech salesperson, the customer support rep may give presentations to the potential customer. However, the customer support rep is primarily concerned with assisting and supporting the sales staff by helping to solve technical problems, assessing feasibility of implementing programs, demonstrating and installing equipment, and helping to develop sales proposals. The customer support rep must be well versed in her company's product line and capabilities.

JO MARTELL Jo Martell is a sales engineer at Rolm Inc., a telecommunications company headquartered in Silicon Valley. She provides technical sales support in Rolm's New York office. After majoring in history and education in college, she wanted to use her educational background in business. Her first job came through an employment agency. Martell began her career at a different telecommunications company. She spent two years in the training department before moving to her present position as sales engineer.

Martell's job requires tremendous attention to detail, an ability to work independently and to give presentations. Continuing education through reading and specialized training courses is vital.

Martell speaks enthusiastically about telecommunications, "I think it's the field of the eighties and nineties. It's interesting, diverse, and it's only going to grow." Rolm's operating style is MBO, management by objective. If its sales exceed quotas each quarter, there are bonuses for employees in addition to yearly salary evaluations. Martell averages fifty to fifty-five hours a week, depending on the sales staff's schedules.

Trainers and Teachers

Education plays a large part in the high-tech world, from the preschool level to continuing education for adults at introductory and advanced levels. This education or training occurs at elementary schools, secondary schools, vocational schools, community and junior colleges, colleges and universities, research and development centers, computer camps, and private computer literacy or data processing centers. Retailers, and manufacturers of hardware, software, and other high-tech products, also provide education, as do many businesses for their employee users.

Teachers may be called EDP (electronic data processing) training specialists, trainers, teachers, and professors depending on the environment in which they work and their field of specialization.

SPECIALTIES Those involved in high-tech education may teach:

- computer literacy
- programming languages and techniques
- computer concepts
- data processing, information sciences, or computer information systems
- computer science
- engineering
- chemistry or physics
- mathematics
- operating skills for various equipment
- technical writing

WHAT AN EDP SPECIALIST DOES A large or medium-sized organization with a data processing division will employ an EDP training specialist or trainer. Her responsibilities will range from orienting new employees and training operators to providing technical courses or seminars about new equipment and techniques for technical employees. A training specialist must:

- evaluate requests for training and assess educational requirements
- identify sources for training
- determine best method of providing training
- meet with organization managers, personnel and finance divisions, to approve programs
- hire and contract instructors; arrange to teach information herself or provide training through arrangement with manufacturer, vendor, or established school
- schedule courses or workshops; arrange for employee attendance
- budget for training

- evaluate training, instructors, programs, and employee response
- determine effects of training program on company as a whole

PRACTICAL VS. THEORETICAL TEACHING At a school, whatever the level of education, a teacher is generally less concerned with business uses and more concerned with transmitting and explaining theory and information. The theoretical emphasis will depend on the level of education. Theory will be covered extensively in graduate and certain undergraduate courses. It will more likely be ignored in vocational courses and computer literacy programs.

An academic environment usually allows for a more informal job structure. However, colleges and universities usually demand advanced credentials for professors (usually a Ph.D.), as well as periodic publications for tenure (job security). The majority of high-tech teachers have direct contact with their students. Some prepare audiovisual material and CAI and other software, but do not themselves enter a classroom.

DEMAND The demand for computer educators is growing by leaps and bounds. The Tandy Corporation, which owns Radio Shack, increased its training staff from 130 to 400 in eighteen months. The number of courses rose from fifteen to thirty-nine. Universities are expanding their degree programs and their continuing education programs. Private computer teaching centers crop up daily. And elementary and secondary schools are adding computer literacy instructors to their faculty. "Opportunities are wide open," says Philip Jones, editor of *Training* magazine.

Backgrounds for computer instructors vary tremendously, including degrees in education, computer science, business, and work experience in other fields. Trainers may move into consulting, sales, marketing, or management.

SALARIES Salaries range from $20,000 to $50,000, with higher salaries at management consulting firms ($30,000 to $100,000).

Consultants

High-tech consultants may be called computer consultants, software consultants, management consultants, industry analysts, financial analysts, computer security specialists, data communications specialists, training specialists, and recruiters or headhunters. Their expertise may be as programmers, systems analysts, software designers, or managers, or in the fields of finance, high-tech education, automation, or telecommunications. The term consultant indicates that work is contracted for by a client and performed in consultation with that client. This position is an independent one; that is, a consultant is hired from outside the company.

WHAT A CONSULTANT DOES A consultant is both a salesperson selling a service based on her knowledge, experience, and judgment, and a problem-solver. Whatever the consultant's specialty, she performs three basic functions for a client:

- diagnoses problems
- recommends solutions
- often helps implement these solutions

DEMAND Typically, the demand for consultants rises with the level of technological or business skills needed for optimum business operation, and with the scarcity of knowledgeable and experienced professionals. With today's explosion of advanced equipment and information processing techniques, good consultants can carve an excellent niche for themselves in the marketplace. High-tech consultants are in demand by companies which have insufficient personnel to handle the complexity of high-tech tasks. (Staff may be inadequately trained or too small to handle the task.) Frequently the required task—such as automating a production line—is considered a one-shot operation that requires an expertise not held by any staff member, and training a staff member or hiring someone new on a permanent basis to accomplish the task would not be cost-effective. By engaging a consultant, an organization secures expertise, experience, less distraction for staff from normal work load, objectivity of an outsider, and the most economical solution.

A consultant, because of her professional independence, is considered the best source for recommendations of hardware, software, information processing and automation techniques, training programs, potential employees, etc. A consultant is paid by the client, not by the manufacturer or vendor of any particular product, nor by any other individual. This helps guarantee her objectivity in proposing solutions.

SELF-EMPLOYMENT VS. STAFF CONSULTING A consultant's independence also indicates the close relationship between this position and the role of an entrepreneur. While many consultants are employed by consulting firms and even by manufacturers and vendors or banks and insurance companies, a large percentage of consultants are self-employed. These consultants cannot be content with providing a service. They must also spend time:

- publicizing and advertising their expertise, availability, and visibility
- unearthing and following leads for possible assignments
- determining desired payment and payment plan
- negotiating payment
- formulating contracts
- consulting with clients
- writing reports
- learning about new products and techniques
- doing administrative work

GAUGING DEMAND FOR SERVICES To be successful a consultant must offer a service that's in demand on a contracted basis —and reach potential clients with the news of how she can solve their problems. If you're a programmer and you'd like to move into consulting, you'll need to determine whether the work you're skilled at (programming in an assembly language?) is contracted out or usually done in-house. (In the case of assembly language programming, the answer is that there's little demand for consultant assembly language programmers.) If there *is* a demand for your expertise—the work is contracted out on a short-term basis—there's a niche for you to fill. For example, if your expertise is an in-depth knowledge of dedicated word processors, the fact that there are so many to choose from points to a strong market for consultants who can analyze a business's needs and operation and determine the best product to satisfy those needs in a cost-effective manner. Because products like word processors can sharply cut a company's expenses—and thus greatly increase profits—there is a wide-open market for consultants in this area.

NEGOTIATING PAYMENT A consulting project may last anywhere from hours to years. A consultant may have an hourly or daily *(per diem)* fee. (This rate must take into account overhead costs and salary.) Or she may charge a fixed amount which covers both *per diem* fees and out-of-pocket expenses. This would be set by contract and is frequently the arrangement made with government agencies. However, this necessitates a clear specification of the consultant's responsibilities and the scope of the assignment—in writing. If the time required to accomplish the task balloons from the initial estimate, the contract should provide for a commensurate adjustment in the fee.

Other fee arrangements include contingency and retainer arrangements. A consultant just starting out might have to prove the benefits of her work by agreeing to a contingency fee—payment to be determined based on the benefits derived from the service performed. This is a risky business and should be entered into with caution only when the consultant is convinced of the results she can achieve and the reliability of the client. It would be preferable to set a lower than desired fee initially, with an agreement to raise it when the client has been satisfied.

Being on retainer for a company means that a percentage of a client's time has been reserved in advance at a preset fee, although the specific work required may not be specified. This offers the consultant regular pay—usually on a monthly basis—and regular work. In return, the consultant is on call to meet the client's needs as they arise and to perform certain periodic services.

PREREQUISITES If you're interested in consulting, consider whether you have sufficient experience, expertise, visibility, sales ability, self-directedness, and financial reserves to carry you through until you get established. (For more on the latter, see Chapter 8.)

SPECIALTIES A high-tech consultant can specialize in such areas as:

- small or large business problems
- personal computers
- word processing
- accounting
- investment analysis
- analysis of industry trends
- high-tech recruiting
- venture capital and high-tech management
- EDP and technical training
- office automation
- factory automation
- data communications
- software design or customizing
- data base administration
- systems analysis
- programming—in a particular language or application
- telecommunications
- computer security

FINANCIAL REWARDS After three years of experience, a consultant might earn a salary well into the twenty- to thirty-thousand dollar range. Salary can rise with experience—and demand—to upwards of $100,000 a year. It is suggested, however, that a would-be consultant begin by moonlighting and holding on to her regular job until contacts are established. Or you might prefer being employed by a consulting firm.

EDP Manager or Director of MIS

This is an upper-level administrative position found in companies with large- or medium-sized data processing or information systems divisions or centers. The holders of these positions maintain overall responsibility for the operation of the computer center, including:

- hiring, assigning, evaluating, counseling, promoting, and firing staff
- relating company policies and goals to staff
- supervising training of staff
- meeting with vendors about new equipment
- evaluating and purchasing equipment
- consulting with other company departments and users about needs and preferences
- consulting with company management to define priorities and to discuss acquisitions and allocation of computer time
- establishing standards
- reviewing project feasibility studies
- scheduling projects and establishing priorities
- maintaining systems and applications programming
- directing and coordinating planning
- writing progress reports for management
- preparing proposals and soliciting sales of services to other companies

PREREQUISITES The EDP Manager or Director of MIS must have a thorough knowledge of the most up-to-date data processing and information systems techniques and equipment and an understanding of how they relate to company needs. These positions require a background in programming or systems analysis, business skills, and managerial experience.

Dr. James Wetherbe, director of the Management Information Systems (MIS) Research Center at the University of Minnesota's School of Management has noted that MIS executives have the greatest management challenges and opportunities in modern organizations. Why? While the costs of labor, raw materials, energy, transportation, etc. are all escalating rapidly, computer technology offers the means of vastly increasing productivity while putting the brakes on costs. Not surprisingly, these challenges are balanced by high risks. (A 1976 *Harvard Business Review* article observed turnover rates ranging from 25 to 50 percent in the 1970s, many of them involuntary.) When MIS results don't match company expectations, the end result is often dismissal of the individual in charge. Although an unsatisfactory state of affairs may be due more to unrealistic company expectations than to professional incompetence, an individual's career may suffer nevertheless.

There is a tremendous shortage of qualified people for these positions. A well-prepared EDP Manager or MIS Director should have an understanding of business politics and organizational structure, as well as technical and managerial experience. As these positions have evolved over the years, technical skills have taken a back seat to managerial skills. A successful EDP or MIS manager must show loyalty to the organization first, rather than focus exclusively on the technology as in the past. Interpersonal and administrative skills are critical.

Chapter 7, "Moving and Advancing," explores these issues in depth. An aspiring EDP Manager would be well-advised to pursue an M.B.A. and sharpen her business and managerial skills. Acquiring experience in project management, planning, and organization can add to your preparation as well.

SALARIES AND EMPLOYERS EDP Managers earn between $24,000 and upwards of $40,000, depending on experience, background, and the size and importance of their departments. These positions can be found in banks, insurance companies, utilities, transportation organizations, government agencies, manufacturers, retailers, and data processing service companies.

Data Base Managers

Data bases have grown tremendously in importance as we've entered the so-called "Information Age." A data base is a collection of information on a given subject, such as:

- stock market figures
- articles on a specific topic
- airline flights, reservation and travel information
- mailing lists
- political contribution lists
- records of financial transactions

WHAT A DATA BASE MANAGER DOES The data base manager or administrator is in charge of organizing the information, keeping it up to date, avoiding unnecessary duplication, establishing information definitions and standards, and ensuring easy access for users. This may mean coordinating needs and activities of different departments or users.

EMPLOYERS AND SALARIES A programmer or systems analyst might become a data base manager. This position is found at private data base companies (like The Source in McLean, Virginia, Dow Jones News/Retrieval in Princeton, New Jersey, and The Information Store in San Francisco), many professional societies and trade associations (for membership information, job banks, etc.), government agencies, publishing and communications companies, and large businesses. Salaries range from $22,000 to $38,000.

Engineers

The choice of specialization for an engineering student and professional is staggering; and the particular specialty is critical in terms of an engineer's training, future, and compensation. Electrical engineers are in great demand now, but many other engineering disciplines are undergoing dramatic shifts in demand. And these fluctuations are not new to the engineering field.

DEMAND The *1982 US Professional Job/Salary Outlook,* prepared by Fox Morris Personnel Consultants of Philadelphia, notes that "graduates with electrical engineering degrees will be most in demand, followed by mechanical engineering, industrial engineering, and chemical engineering—particularly if they happen to be females, who will account for an estimated 13 percent of the engineering graduate class of 1982."

Electrical engineers may work on computers, medical equipment, missile guidance systems, or power distribution systems.

High-tech experience with synthetic fuels, biotechnology, and computer design are in particular demand. And engineers with backgrounds in finance and management are considered highly competitive in the job market. In addition, engineers interested in teaching will find an abundance of openings, but at salary levels significantly lower than that of their peers in industry.

Aeronautical engineers will find demand keyed to the economic conditions in the airline industry and the size of the defense appropriations.

WHAT AN ENGINEER DOES Engineers may work in research and development, testing, production, operations, maintenance, sales, or customer support. They often work closely with computers and calculators on simulations. Engineers may be grouped together for complex projects or work individually on smaller projects. Some senior engineers supervise less experienced engineers and have administrative responsibilities for budgeting, scheduling, decision-making, and personnel matters.

PREREQUISITES Engineers must be technically proficient and able to answer questions regarding feasibility, and more frequently today, questions of cost-effectiveness. Engineers must understand the basic theories and principles of science and mathematics and then apply them to practical technical problems. They may design products (about 50 percent of all engineers are employed in manufacturing), systems, or processes, and work on their development or implementation.

Engineers may have undergraduate or graduate degrees. They

must be creative, analytical, and detail-oriented. Good oral and written communication skills are also necessary. Continuing education is crucial for an engineer's professional success.

SALARIES A recent College Placement Council survey concluded that engineering graduates receive higher salary offers than all other professionals in the United States. In 1981, the average salary for engineers with undergraduate degrees but no experience averaged $22,900 in private industry. A master's degree raised the salary to $25,500 annually; a Ph.D., to $32,800. Salaries vary according to the branch of engineering. Petroleum engineering offers the highest salaries; aeronautical and civil engineering the lowest.

A FINAL WORD

The previous pages have explored the major high-tech positions in demand today. Yet this chapter has only hinted at the large numbers of employment opportunities available for those who want to *use* the latest technology in positions not generally considered high-tech in themselves. You might want to pursue a career as a computer graphics artist, an art conservator, an astronaut, a financial analyst, a politician, a stockbroker, a librarian, a doctor, a medical technician, a fashion designer, a farmer, a lawyer, an accountant, or a pilot. The following chapter will help you decide which is the best job for you.

5

Getting the Right Job for You

Having accepted the fact that the high-tech world offers you exceptional professional growth in what may well be the marketplace of the 1980s and beyond, how do you go about choosing the job that's right for you? (Yes, getting a job is more than a matter of assessing your wants and needs, your interests and aptitudes, your values and personality, and your training or education.) The demands of the marketplace are explored in Chapters 1 and 4; job search techniques in Chapter 6. In this chapter, you'll work your way toward identifying the best job for you, and have fun along the way.

Following are a number of quizzes based on the latest research concerning aptitudes, interests, values, and personality characteristics, and how they relate to specific high-tech positions. Take your time answering these questions. The more accurate your answers, the better the match will be between you and suitable positions.

Keep in mind that it's not uncommon for individuals who don't fit the usual pattern of interests and values, even of personality characteristics and aptitudes, for a particular profession to nevertheless be quite happy and successful in that profession. The self-surveys that follow can help you pin down what's important to you—and how those likes or dislikes are satisfied by particular jobs.

These tests help you select options. Do not use them to shut off opportunities that interest you. Satisfying careers typically involve an element of risk-taking. If you're interested in programming, but your score indicates a less than perfect match, proceed—with caution. You may find that teaching programming is a better choice. Yet two personality characteristics can override a whole slew of seeming mismatches: stubbornness, a.k.a. dedication or commitment, and hard work. Use these tests as a guide, but recognize that you may well know yourself best of all.

The positions explored in this chapter include those of:

- computer service technician
- programmer
- systems analyst
- technical writer
- salesperson
- customer support representative
- teacher or trainer
- consultant
- EDP manager or director of MIS
- data base manager
- engineer
- financial analyst
- recruiter
- computer graphics artist

Of course, there are numerous other high-tech positions, but these are the major ones.

Ready to look deeply into your psyche? O.K., here we go.

Test One

1. Are you gifted at abstract thinking?
2. Are you conscientious and persistent?
3. Are you self-confident and self-assured?
4. Do you consider yourself assertive and dominant?
5. Are you adaptable and flexible?
6. Are you trusting?

7. Are you cool and reserved?
8. Are you warm and spontaneous?
9. Are you self-disciplined?
10. Do you like to experiment, to try new things?
11. Are you easily angered or upset?
12. Is your thought process a logical one?
13. Do you like puzzles?
14. Do you like to play chess?
15. Are you curious about how things are put together?
16. Do you have mechanical ability?
17. Are you patient with yourself and with others?
18. Do you like to teach?
19. Can you be persuasive?
20. Are you analytical?
21. Do you consider yourself creative?
22. Can you follow rules exactly?
23. Do you learn quickly?
24. Do you embrace change easily?
25. Do you like to work independently?
26. Do you enjoy working with people?
27. Do you have leadership abilities?
28. Do you like to work in spurts?
29. Do you prefer steady work?
30. Are you organized?
31. Are you articulate?
32. Do you listen well?
33. Do you like to work with things?
34. Do you like detail work?
35. Do you prefer to think in terms of concepts and principles?
36. Do you mind interruptions?
37. Are you gregarious?
38. Are you shy?
39. Are you orderly?
40. Are you a perfectionist?
41. Do you take criticism easily?

42. Are you tolerant of others?
43. Do you prefer to rely on your own knowledge?
44. Are you full of ideas?
45. Can you plan ahead?
46. Can you make decisions easily?
47. Can you weigh pros and cons?
48. Do you like working with your hands?
49. Do you prefer using your head?
50. Do you have a good memory?

Phew! Fifty questions and that's only Test One, you exclaim. Let's take a breather to talk about how answering these questions helps reveal the best career direction for you.

TIME OUT FOR A SPOONFUL OF THEORY

Researchers interested in psychological testing of high-tech professionals are a varied group. There are industrial psychologists, high-tech company personnel staff, organizational psychologists, data processing professionals, consultants, management analysts, guidance counselors, non-profit research organizations, and career counselors and testing services. Their goals are the same: to find out what makes successful and happy high-tech employees. Along the way, researchers have made interesting connections that help contribute to a profile of the typical high-tech professional. But more on that later.

Now look at your answers to Test One. Count your yes answers. The following key indicates your general career direction:

Key

0-19: This score uncovers a touch of creativity, which might qualify you for a position as a writer or artist or even recruiter.

20-28: This score suggests your interest in working with people. Positions to consider include those of customer support representative, salesperson, teacher or trainer, Director of MIS or EDP manager.

29 or over: This score reveals a technical or independent bent. You might be happy as a computer service technician, programmer, systems analyst, or consultant.

Ready for Test Two?

Test Two

1. Can you admit your mistakes?
2. Are you easily frustrated?
3. Are you responsible?
4. Are you diplomatic?
5. Is your style direct?
6. Are you energetic?
7. Do you have a sense of humor?
8. Are you ambitious?
9. Do you manage your time well?
10. Do you want to be highly visible?
11. Is self-expression important to you?
12. Do you like to change your surroundings frequently?
13. Do you thrive in a chaotic work atmosphere?
14. Do you prefer to work at home?
15. Do you work well under pressure?
16. Are you competitive?
17. Do you like to play political games?
18. Do you like to take risks?
19. Is job security very important to you?
20. Do you like to take the initiative?
21. Is it important to you to be liked?
22. Is it most important to you to be respected?
23. Do you relate well to men?

24. Do you relate well to women?
25. Do you prefer a structured work environment?
26. Are you friendly by nature?
27. Do you honestly like people?
28. Are you interested in attaining power?
29. Is self-fulfillment important to you?
30. Are you a nurturer?
31. Are short-term plans more important to you than long-term plans?
32. Do you like to calculate and quantify things?
33. Do you like to help others, be of service?
34. Do you like to influence others?
35. Do you like to help others learn and grow?
36. Do you like to research new things?
37. Do you enjoy observing and analyzing people and how they interact?
38. Do you enjoy observing and analyzing events?
39. Do you enjoy observing and analyzing things?
40. Do you like to invent or create new things?
41. Do you like to see a project through from start to finish?
42. Do you like constant contact with other people?
43. Do you like your work to be routine?
44. Do you prefer a flexible work arrangement?
45. Is your work the most important part of your life?
46. Are you willing to have your work and personal lives overlap?
47. Can you accept being in the minority?
48. Are you willing to be a trailblazer?
49. Is your physical appearance important to you?
50. Is your style of dress important to you?

TIME OUT

You're progressing nicely—and getting closer to the pay-off: a better understanding of which high-tech positions would prove most satisfying to you. By this point you may be curious about how such issues as your feelings about your physical appearance relate to the best job for you. After all, you're not applying for a position as a model or an actress. The answer is that some positions require your getting your hands dirty—and often the rest of you as well. If you're the type who's uncomfortable whenever a hair is out of place, you'll have to rule out those positions or make peace with temporary disorder.

Count your yes answers to Test Two, then read below:

Key

0-10: This score indicates your technical leanings. Consider a career as computer service technician or programmer.

11-25: A score in this range signals creativity. You would not be satisfied in a purely technical position.

26 or over: You are people-oriented. Keep this in mind when you select a career that will satisfy you. (Consider sales and education.)

Ready for Test Three? Your move.

Test Three

1. Do you have high standards for others?
2. Do you maintain high standards for yourself?
3. Are you easily frustrated?
4. Are you challenged by problems?

5. Can you cope with disorder?
6. Do you feel inadequate?
7. Are you opinionated?
8. Do you consider yourself an enthusiastic person?
9. Are you inclined to be casual?
10. Do you enjoy being dramatic?
11. Are you optimistic in outlook?
12. Are you a brooder?
13. Are you a procrastinator by nature?
14. Do you like to "wing" things?
15. Is always knowing all the answers important to you?
16. Are you willing to admit ignorance on occasion?
17. Have your happiest moments been spent with others?
18. Can you set priorities?
19. Are you obsessed with details?
20. Do people fascinate you?
21. Are you easily sidetracked?
22. Are you conscientious?
23. Can you handle failure?
24. Is it important that you continually prove yourself?
25. Are you tactful?
26. Are you sensitive?
27. Are you thin-skinned?
28. Do you dislike confrontations and avoid them at all costs?
29. Do you like to play peacemaker?
30. Do you prefer groups to one-on-one interaction?
31. Do you lose your temper frequently?
32. Do you believe in honesty on all occasions?
33. Are you basically a loner?
34. Are you tenacious?
35. Do you consider yourself a well-rounded individual?
36. Do you see the world as a hostile place?
37. Are personal relationships important to you?
38. Do you have a forceful personality?
39. Do you know how to give in gracefully or admit defeat?

40. Do you prefer diversity to a status quo?
41. Are you future-oriented?
42. Are you neat and orderly?
43. Are you humble?
44. Are you precise?
45. Are you usually serious in manner?
46. Are you reserved about your emotions and personal life?
47. Is it more natural for you to enjoy others than to judge them?
48. Are you straightforward and down-to-earth?
49. Do you worry about past mistakes?
50. Is the approval of others important to you?

TIME OUT

By now you may be thinking that identifying the position that's right for you is hard work. It does require time and effort to find your match, but when you consider that the average worker spends 80,000 odd hours at her chosen profession, the frustrations of a poor job match seem no small thing.

There are career counselors who "hold your hand" as they take you through the maze this chapter is traveling through. And there are well-respected non-profit research organizations, like the Johnson O'Connor Research Foundation, who help individuals and companies test aptitudes, and then provide counsel on the basis of their findings. In addition, there's the service provided by the Educational Testing Service (ETS) of Princeton, New Jersey, which offers a computer-based career assistance program. *SIGI,* System of Interactive Guidance and Information, asks questions and processes answers to identify occupations which best meet an individual's need for prestige, independence, helping others, security, variety, leadership, interests, leisure time needs, income level preferences, and career preparation time. SIGI is based on information concerning 300 occupations, and takes into account work activities, prerequisites, income, personal satisfactions, working conditions, and outlook for job opportunities of the occupation. Lastly, SIGI assists individuals in evaluating occupations by comparing risks and rewards—doing a cost-benefit analysis. SIGI is presently available—in a three-hour program—at nearly 300 colleges and universities. It is generally free of charge for students, and occasionally available for use by members of the community as well. New occupations titles are added as warranted. (One of the more recent additions includes robotics technician.) For more information, write SIGI, Educational Testing Service, Rosedale Road, Princeton, New Jersey 08541.

Let's examine your answers to Test Three. Again, count your yes answers.

Key

19 or under: You value your independence. You'd do best in positions where you are your own boss. You might even have the makings of an entrepreneur; at the least, of a consultant or free-lance writer or artist.

20 or over: Independence is less important to you than contact with people. You value feedback and look to others for support and personal satisfaction. A position as teacher, salesperson, or even systems analyst might be a good choice for you.

You're in the home stretch now. Here's the final exam, Test Four.

Test Four

1. Are you concerned about the protection of your privacy?
2. Is autonomy important to you?

3. Do you need to be challenged by your work?
4. Is earning a high income important to you?
5. Is earning a stable salary important to you?
6. Are you concerned about data security?
7. Is job security critical to you?
8. Are you anxious about dealing with mathematics?
9. Do you embrace new technology?
10. Do you admit ignorance or ask questions easily?
11. Do you enjoy playing video games?
12. Do you think the information revolution will close doors?
13. Do you believe in the value of computer education?
14. Do you know successful women in the field of your choice?
15. Are you comfortable making compromises to achieve your ends?
16. Do you prefer tranquillity?
17. Are you single-minded?
18. Do you know yourself well?
19. Do you have a tendency toward defensiveness?
20. Do you like to be left alone sometimes?
21. Can you tolerate fools?
22. Do you have a financial reserve to fall back on?
23. Are you a perfectionist?
24. Do you prefer being master of your own fate?
25. Do you feel it's necessary to prove your worth to others?
26. Do you know how to network?
27. Are you able to accept emotional support from others?
28. Can you meet deadlines?
29. Can you present a professional image?
30. Can you play the corporate game?
31. Are you willing to play company politics to get ahead?
32. Do you resent being judged solely by the quality of your work?
33. Do you like public speaking?
34. Can you toot your own horn?

35. Do you know how to negotiate?
36. Are you generally in control?
37. Are you a hustler?
38. Do you want to become an expert?
39. Are you comfortable talking about money?
40. Do you know what you're worth?
41. Are you comfortable dealing with all levels of employees?
42. Do you like working with children?
43. Do you prefer working with adults?
44. Do you prefer nicely furnished, classy surroundings?
45. Do you need a room of your own to work well?
46. Do you know how to address different individuals, different audiences?
47. Do you know how to do your own research?
48. Do you like being anonymous at work, a cog in a big wheel?
49. Is camaraderie at work important to you?
50. Is your self-esteem tied up with your professional success or failure?

TEST OVER

In addition to SIGI, ETS will offer a new career guidance and information system for adult career changers. This interactive computer service will be available in the fall of 1985. There are currently two other computer-based career guidance systems available. DISCOVER, from the American College Testing Service, is largely directed to college students. It includes information on 425 occupations, 1,481 two-year colleges, 1,761 four-year colleges, and 1,021 graduate schools. CHOICES, offered by the Virginia-based Canada Systems Group (CSG) is designed to help guide high school students in their career choices.

Tally up your positive answers to Test Four.

Key

0-14: This score indicates a strong interest in helping or teaching others. You should consider a position as teacher, trainer, or customer support rep.

15-27: Your primary professional interest is in problem-solving—repairing, inventing, adapting, improving, and explaining technology. Think about a position as computer service technician, programmer, systems analyst, technical writer, salesperson, EDP manager, director of MIS, financial analyst, or computer graphics artist.

28 and over: You have a drive for independence that might be best satisfied in the fields of consulting or recruiting. Or consider the rewards—and the demands—of the entrepreneurial life.

WHAT'S IT ALL ABOUT?

Tests One through Four have been designed to suggest your career aptitudes and desires. Using the keys that follow each test, you've identified your primary leaning as technical, people-oriented, or independent. Now you're ready to hold up all your answers against the profiles of each high-tech position. Remember that an exact fit is unlikely. See how close your answers come to the profiles of the positions you're interested in. Then you might want to turn back to the previous chapter to learn more about the job you're interested in. The following chapter will equip you with the skills you need to locate and land your chosen high-tech job.

Keep in mind while you compare your answers to those of the profiles below that only the yes answers count, and only the yes answers listed for each test are really significant. This means that if you've answered yes to other questions in addition to the ones listed below, you may still have found a good match. Those numbers with an * signify traits which are considered critical for success in the particular position.

In general, these positions can be considered primarily technical: computer service technician, programmer, systems analyst, engineer, and financial analyst. More people-oriented positions are technical writer, salesperson, customer support rep, teacher/trainer, consultant, EDP manager/director of MIS, data base manager, and recruiter. Positions that appeal to independent-minded individuals include consultant, recruiter, and free-lance technical writer or artist. These same positions may also be appealing to those creatively inclined.

COMPUTER SERVICE TECHNICIAN

Test One

Yes: 1, 2, 5, 7, 9, 10, 12, 13, 15, 16, 17, 20, 21, 22, 23, 24, 25*, 29, 30, 32, 33, 34*, 38, 39*, 40, 41, 44, 47, 48*, 50.

Test Two

Yes: 1, 12, 19, 23, 31, 32, 33, 39*, 47, 48.

Test Three

Yes: 2, 3, 4, 5, 6, 9, 12, 19, 22, 24, 28, 30, 32, 33, 34, 42, 44.

Test Four

Yes: 5, 7, 9, 10, 17, 20, 23, 28, 29, 36, 38, 43, 47, 48.

PROGRAMMER

Test One

Yes: 1, 2, 5, 7, 9, 10, 12, 13, 14, 17, 20, 21, 22, 23, 24, 25*, 29, 30, 32, 34*, 36, 38, 39*, 40, 41, 43, 44, 47, 50.

Test Two

Yes: 1, 7, 15, 23, 31, 39, 40, 47, 48.

Test Three

Yes: 2, 4, 5, 6, 9, 12, 19, 22, 24, 30, 32, 33, 34, 42, 44.

Test Four

Yes: 9, 10, 13, 17, 20, 23, 28, 29, 38, 43, 46, 47, 48.

SYSTEMS ANALYST

Test One
Yes: 1, 2, 5, 7, 9, 10, 12, 13, 14, 17, 19, 20, 21, 23, 24, 26, 30, 31, 32, 34, 35, 38, 39, 44, 45, 46, 47*, 49, 50.
Test Two
Yes: 1, 4, 5, 6, 7, 12, 13, 15, 20, 22, 23, 36, 39, 40, 41, 42, 45, 47, 48.
Test Three
Yes: 1, 2, 4*, 5, 6, 11, 12, 15, 17, 18, 19, 22, 24, 26, 34, 38, 39, 40, 41, 42, 44, 49, 50
Test Four
Yes: 2, 4, 6, 9, 10, 20, 28, 29, 30, 31, 32, 35, 36, 38, 41, 43, 46, 47, 50.

TECHNICAL WRITER

Test One
Yes: 8, 9, 10, 11, 19, 20, 21*, 22, 23, 24, 25, 30, 31, 36, 38, 42, 44, 49*.
Test Two
Yes: 5, 7, 9, 14, 19, 29, 33, 36, 38, 39, 40, 41, 44, 47, 48.
Test Three
Yes: 4, 8, 9, 10, 12, 16, 17, 18, 19, 20, 22, 24, 26, 27, 29, 34, 35, 37, 41, 42, 44, 47, 48, 49.
Test Four
Yes: 2, 5, 9, 10, 20, 25, 28, 29, 41, 43, 46*, 47*, 48, 50.

SALESPERSON

Test One
Yes: 2*, 3*, 4*, 5, 8, 9*, 10, 17, 19*, 20, 21, 23, 24, 26*, 30, 31*, 32, 37*, 41, 42, 44*, 45, 49, 50.

Test Two
Yes: 3, 4*, 6*, 7*, 9, 11, 12, 18, 21, 23, 26, 27, 33, 34, 37*, 38, 42*, 44*, 47, 48, 49, 50.
Test Three
Yes: 4, 8, 10, 11, 14, 16, 17, 18, 20, 22, 23, 24, 25, 26, 29, 34, 35, 37, 38, 40, 47, 48, 50.
Test Four
Yes: 2, 4, 9, 10, 15, 18, 20, 21, 24, 26, 29, 33, 37, 39*, 40, 43, 44, 46*, 47, 50.

CUSTOMER SUPPORT REP

Test One
Yes: 5, 8, 12, 15, 16, 17, 18, 19, 20, 22, 23, 24, 26*, 28, 30, 31*, 32, 37, 42, 44, 46, 50.
Test Two
Yes: 3, 4*, 7, 19, 21, 23, 30, 33*, 39, 42*, 47, 48.
Test Three
Yes: 4, 8, 11, 17, 20, 22, 24, 25, 26, 28, 29, 37, 43, 44, 47, 48, 50.
Test Four
Yes: 5, 20, 21, 29, 32, 38, 43, 48.

TEACHER/TRAINER

Test One
Yes: 3, 10, 17, 18, 19, 21, 24, 26*, 30, 31*, 32, 35, 37, 42*, 43, 44*, 45, 46, 49, 50.
Test Two
Yes: 3, 4, 5, 6*, 7, 9, 10, 11, 20, 23, 27, 29, 30*, 33*, 34*, 35*, 36, 37*, 39, 42*, 44.
Test Three
Yes: 7, 8, 9, 10, 14, 15, 16, 17, 18, 20, 22, 23, 25, 26, 28, 29, 30, 35, 37, 38, 44, 47, 48, 49.

Test Four

Yes: 5, 13, 20, 21, 29, 32, 36, 38, 42, 43, 47.

CONSULTANT

Test One

Yes: 1, 2, 3*, 4, 5*, 8, 9*, 10*, 13, 18, 19*, 20, 21, 23, 24*, 26*, 28*, 30*, 31*, 32*, 37, 42, 43, 44*, 45*, 46*, 47*, 50*.

Test Two

Yes: 3, 4*, 6*, 7*, 8, 12*, 15, 16, 18*, 20*, 23, 36*, 39, 40, 41, 42, 44*, 45, 46, 47, 48, 49, 50.

Test Three

Yes: 2, 4*, 5, 11, 18*, 22, 23, 24, 25, 26, 34, 35, 38, 40, 44, 48, 50.

Test Four

Yes: 2, 4, 6, 15, 20, 21, 22, 24, 25, 26, 27, 28, 29, 32, 33, 34, 35, 36, 37, 38, 39*, 40*, 41, 43, 46, 47*, 50.

EDP MANAGER/DIRECTOR OF MIS

Test One

Yes: 1, 2, 3, 4, 5, 9, 10, 18, 19, 20*, 23, 24*, 26*, 27*, 30*, 31, 32, 39, 42*, 45*, 46*, 47*, 49, 50*.

Test Two

Yes: 1, 3, 4, 5, 7, 8, 9, 10, 13, 17, 20*, 22, 23, 28, 34, 36, 37, 39*, 42, 45, 47, 48.

Test Three

Yes: 2, 5, 15, 18*, 20, 22, 23, 24, 25, 26, 29, 30, 38*, 40, 41, 44, 48.

Test Four

Yes: 4, 6, 10, 15, 20, 29, 30, 33, 35, 36, 41, 43, 44, 45, 46.

DATA BASE MANAGER

Test One
Yes: 1, 2, 9, 10, 12, 13, 20, 24, 29, 30*, 39, 43, 50.
Test Two
Yes: 19, 22, 23, 40, 41, 43, 47.
Test Three
Yes: 2, 15, 17, 18, 19, 22, 24, 30, 42, 44, 48.
Test Four
Yes: 1, 4, 17, 21, 29, 38, 41, 43.

ENGINEER

Test One
Yes: 1, 2, 5, 9, 10, 12, 13, 15, 16, 17, 20*, 21, 24, 33, 34, 38, 39, 43, 47, 48, 50.
Test Two
Yes: 1, 22, 23, 36, 39, 40, 41, 47, 48.
Test Three
Yes: 2, 4, 5, 6, 12, 16, 17, 18, 19, 22, 24, 30, 32, 33, 34, 42, 44, 48.
Test Four
yes: 2, 20, 28, 29, 36, 38, 43, 47.

FINANCIAL ANALYST

Test One
Yes: 7, 9, 10, 12*, 13, 18, 19, 20*, 22, 29, 30*, 34*, 38, 39*, 40, 43, 45, 46, 49.
Test Two
Yes: 1, 4, 22, 23, 25, 32, 39, 41, 43, 47, 48.
Test Three
Yes: 2, 4, 15, 19, 22, 24, 25, 32, 41, 42*, 43, 44*, 48, 49, 50.

Test Four
Yes: 1, 2, 4, 20, 21, 23, 29, 35, 36, 38, 39*, 40*, 43, 44, 45, 47.

RECRUITER/HEADHUNTER

Test One
Yes: 2*, 3*, 4, 5, 9*, 10, 19*, 23, 26*, 28*, 31*, 32*, 37, 44*, 45, 46, 50*.
Test Two
Yes: 3, 4, 5, 6*, 7, 16, 17, 18, 20, 32, 34, 36*, 37, 41, 42, 44, 46, 47, 48, 49, 50.
Test Three
Yes: 1, 8, 10, 11, 17, 18*, 20, 21, 22, 23, 24, 25, 28, 34*, 38*, 39, 40, 48, 49, 50.
Test Four
Yes: 1, 2, 4, 6, 9, 10, 15, 17, 20, 25, 26, 28, 29, 33, 34, 35*, 37, 38, 39*, 40*, 41, 43, 44, 46, 47*, 50.

COMPUTER GRAPHIC ARTIST

Test One
Yes: 8, 9, 10, 11, 21*, 22, 24, 25, 33, 34*, 38, 39, 40, 44.
Test Two
Yes: 9, 11, 29, 31, 39, 40, 41, 47, 48.
Test Three
Yes: 2, 4, 9, 10, 14, 21, 22, 24, 26, 27, 35, 37, 40, 44, 49, 50.
Test Four
Yes: 2, 11, 20, 22, 23, 25, 27, 28, 34, 43, 46, 50.

Getting Your Foot in the Door

What kind of first job you set your sights on will depend on factors like your educational background, interests, aptitudes, skills, goals, personality, desired work environment, and preferred geographical location.

The preceding chapter has helped you identify specific positions that might suit you. This chapter will focus on the logistics of identifying employment opportunities, presenting yourself for consideration, and negotiating a job offer. How you present yourself—in your résumé, in letters, in telephone calls, and in person—is critical to your success. Securing your first position is essentially a two-step process:

1. discovering that a suitable position exists—*or can be created,* and
2. convincing potential employers that you are the right person for the job.

It sounds simple when put this way, but don't be fooled. Job hunts are often life's toughest and most exhausting journeys. They can be more traumatic than illness and accidents, especially for the neophyte job hunter. You'll experience the exciting highs of landing interviews, and the dismaying lows of being passed over for a position. It's all part of the process where a positive and realistic attitude (self-confidence blended with modesty), hard work, discipline, and imagination will pay off. Don't lose sight of your own personal worth, and of the fact that a high-tech job does indeed exist for you.

A glance at the dark side of the job hunt is a good idea before proceeding with your search. Optimism and enthusiasm make for excellent ammunition in the "battle," but so too does the recognition that rejections are all too common. You can't win every time, but continuing efforts will help refine your presentation skills and bring you closer to landing your first job.

This chapter will help you learn to:

- identify job opportunities through a multitude of sources, both formal and informal
- understand what really goes on during the course of a job interview—what to expect and how to prepare, how to answer questions *and* what questions to ask and how to ask them, what information to volunteer, how to dress, and how to listen
- know when to discuss salary and how to negotiate
- handle rejection and the brighter side of the coin, acceptance
- analyze a particular position and a company's suitability—in terms of risks, challenges, learning opportunities, job mobility, special benefits and drawbacks, management and company style, congeniality and support of co-workers and managers
- acquire a complete picture of a position's responsibilities, activities, necessary skills, and special requirements
- understand company policies regarding evaluations, promotions, raises, the status of entry-level workers and of women, and the fate of previous job holders.

SPOTLIGHTING YOUR ACCOMPLISHMENTS

What do you have to offer an employer? If you're fresh out of school, the question may prompt a severe crisis in confidence as you face the business world—perhaps for the first time—and fear your inadequacy. You may feel adrift in a foreign environment because the language of business—and of job hunting in particular—may be new to you. You may fear you have nothing worthwhile to offer because you haven't learned yet how to frame your strengths in business terms.

If you've excelled in school, you may expect to be greeted with the same high praise on an interview or in a job. You may have been led to believe that people with your training are in great demand. And so you may approach your job hunt with the erroneous—and self-destructive—attitude that you needn't prove yourself anew. Nevertheless in today's job market you'll face competition for almost any high-tech job you seek *at the entry level.* And while self-fulfillment may be *your* most important goal, a company's representatives will be far more interested in your knowledge of the company and your analysis of how you can contribute to *its* goals.

Remember too that modesty does have a place in your self-presentation. In fact, one of the gravest mistakes you can make on a job interview is presuming to tell potential employers how to run their businesses. What you learn in school does not always correspond with actual business practice. There are factors such as cost and profitability, politics, government regulations, and more to explain a particular state of affairs. Be prepared to listen and learn.

A Chicago-based outplacement company, Challenger, Gray & Christmas, which counsels managers on job search strategies, advises job seekers to concentrate on convincing potential employers of the excellent job they can do for the company. Offering unasked for advice on an interview can get you, in effect, kicked out the door.

A newly-minted graduate may well be considered a treasure by a high-tech company dedicated to training employees its own way, or by a company that seeks to establish company loyalty right from the start. John Imlay, chairman of MSA, an Atlanta-based software company, explains his preference for new graduates: "We like to train them in our way," he says, echoing a not uncommon point of view.

Larger companies like Hewlett-Packard have extensive college recruiting programs, because they believe in them as a source of excellent entry-level employees. (It's worth noting that the existence of a college recruiting program often suggests a receptive company attitude toward providing tuition-remission for further education and frequently toward providing in-house training as well.)

If you've taken advantage of internship, co-op, or apprenticeship programs while in school, you should highlight these experiences in your presentation. If you've held part-time jobs working with computers—programming, selling, instructing, operating, writing about, repairing, or even designing them—play up your experience. And if you have computer know-how from coursework in computer science, data processing, information systems, business, engineering, or other subjects, by all means be specific about that too.

Unrelated job experience can indicate your success at operating in business environments. Good grades will be taken as a sign of your ability to work hard and master knowledge. Be modest, but don't underrate yourself. Self-deprecation can be charming occasionally, but it is more likely to be taken seriously as a sign of low self-esteem in a job applicant. On a job hunt, you have to be your number one fan. So keep your strengths in the forefront of your mind.

RECENT GRADUATE STEREOTYPES

If you lack experience, be mindful that your recent-graduate status may signify many valuable pluses to a potential employer, such as:

- knowledge of the latest developments in your industry, at least in theory
- willingness to start low and work up

- appreciation of education and willingness to take on additional training as needed
- willingness to relocate
- ambition
- eagerness
- high energy level

It may not matter whether these are really *your* personal attributes. You may still want to play up these characteristics whenever your lack of experience is mentioned. Let a stereotype work for you.

On the other hand, recent graduates may suffer from unspoken *biases* against them. They may be thought to be:

- immature—lacking in interpersonal skills
- unstable—likely to job hop
- overly ambitious—expect too much too soon
- unreliable
- in need of too much attention, training
- lacking in company loyalty
- not sufficiently concerned with representing the company well
- unlikely to recognize the importance of professionalism
- expectant of special treatment, consideration

Again, how *you* measure up against this negative profile may mean little if a potential employer believes in this stereotype and won't give you time to prove yourself. However, an awareness of this all too common negative stereotype can help you mount a frontal attack. Focus on those of your personal strengths which counter the negative charges that may be lodged against you.

REENTRY WOMAN

If you're looking for your first job after time spent caring for a family, you may also suffer from unspoken biases—and benefit from others. You may have to fight against such misconceptions as:

• you'll be grateful for any job
• you'll take less money because you don't know your own worth
• you won't be interested in promotion—or relocation
• you won't be able to work overtime when necessary
• you'll have less energy than a younger worker
• you'll "mother" everyone in the office
• you'll have a difficult time adjusting to a business environment
• your priorities won't be your work
• you'll have low self-esteem
• you'll be inflexible and rigid

Or you may benefit from the stereotypical advantages reentry women are said to enjoy:

• maturity
• reliability and stability
• life experience
• planning and management skills
• self-confidence
• work commitment

It's no accident that some opposite traits appear on each list. While you as an individual may bear some of the characteristics on each list, a biased employer will decide in advance that they are assets or debits. Again, capitalize on your strengths.

JOB OR CAREER CHANGERS

If you've been working in another profession and want to enter the high-tech world, this may mean taking an initial pay cut —or sizably increasing your salary, depending on your present position. If you have no background in computers, you may have to begin as a trainee. Make sure that the job you select fits into a viable career path and is not a dead end—seek out, for example, applications programming over data entry (a notorious dead end). If you use word processing equipment in your present job, you might consider a move into the more lucrative and upwardly mobile field of training users or selling the equipment. If you've been teaching high school, you might consider a natural move into computer or software training in business or education, or a position in customer support.

You might want to redirect your career from repairing radios and televisions to repairing mainframes or personal computers. Or you could make better use of your selling talents in computer sales than in stationery supplies.

Whether you need formal retraining or not will depend on your background and on the particular job you want to pursue. Many people with backgrounds in math, engineering, physics, chemistry, business, education, foreign languages, and even music have successfully made the transition to high-tech careers. So have individuals with other backgrounds. You'll need to identify the high-tech positions you aspire to (with the assistance of Chapter 5), and plan a careful strategy to reach your goal.

First look within your present company. Is there a high-tech division or position you can move to? Will your present employer train you? Or will the company reimburse you for high-tech training you pursue at local schools or universities? Can you prove your potential by taking an aptitude test? Have you revealed your interest and enthusiasm to those who have the power to help you reach your goal? To managers who could hire you? To personnel staff who could notify you of openings, arrange tests and interviews?

If there's nothing promising at your present company, you'll have to follow a path similar to that of the first-time job seeker—with the twist of playing up your applicable experience and maturity. An employer may hold a narrow view of the background and qualifications needed for a particular position. It's your task to demonstrate the relevance and importance of the skills you have, how you can benefit the company with your:

- communication skills—as a member of a team you must work effectively with others, speaking and writing clearly and listening well
- problem-solving ability—more significant than particular technical skills is the ability to put it all together to propose and implement solutions
- organizational skills—in a high-tech career, the ability to set and follow priorities in a high-pressure environment is at a premium
- flexibility—being able to adapt well to change (new products, new developments, new services, new methods of operation) is critical

Too often career changers underestimate the value of their experience and skills, and fail to see the transferability of their backgrounds for high-tech careers. Whatever your background, there *is* a high-tech job for you, one that builds on previously acquired skills and past experiences, and offers promise for future professional growth.

LOCATION

Which geographical areas have the highest demand for entry-level high-tech workers? Should you consider moving before you line up your first job? Do high-tech jobs exist in all locations? Are you better off in a large city or in a small town or rural area? Does Silicon Valley hold a corner on the high-tech job market? Where are high-tech salaries highest? Where is the cost of living lowest?

These are all important questions to consider when weighing the merits of one location against that of another as you prepare to search for your first high-tech job. Uprooting yourself without a job to call your own is a risky, nerve-racking experience. You would need sufficient financial and emotional resources to tackle a job hunt successfully under those circumstances. It can be done, but it's far from the easiest way.

Do you want to remain in your general area? How far away are you willing to relocate? Do you prefer cities to small towns? Where do you have friends or relatives who can offer you support during your job hunt? How independent are you? How quickly do you acclimate yourself to new surroundings, make new friends? Would you be happy in the northeast? The south? The west? The midwest? Another country? Consider your answers carefully. How much are you willing to sacrifice in location for your career?

The geographical area that offers you the greatest high-tech opportunity will depend on the particular industry you're interested in, the economic conditions of the area at the time, the competition for jobs in the area, the desirability and popularity of the area, the number and quality of its advanced educational institutions, and its cost of living.

WHAT FIRST JOB?

The best-known entry-level job in high-tech companies is the programmer-trainee or junior programmer position. There is more competition for these slots than ever before, which makes your presentation more critical than ever. When you're competing with others who have the same basic training as you do—or better—how you present yourself can spell the difference between landing the job and losing out. Let's examine the programming situation further.

Though the need for computer professionals has been growing apace, newly trained programmers without experience often face difficulty securing appropriate jobs today. Many companies have cut down on the number of entry-level programmers they hire, while other employers have become more selective. In the case of programmers, the number and type of computer languages a programmer knows have become increasingly important. So too has the type of equipment used—mainframe, micro- or minicomputer, brand—and specific software, if used. A preference for higher degrees may also prevail, with graduate degrees preferred to undergraduate ones, four-year degrees winning out over two-year programs, and two-year programs being accepted over training of six months or less. And programming skills alone may not be considered sufficient. Employers may want to see business skills—problem-solving ability and good communication skills—as well.

Some companies have turned to training programmers in-house, thus reducing their need for entry-level programmers from outside the company. Others are relying more on pre-existing software packages instead of on programmers, or adapting commercial software to their own purposes—eliminating or severely diminishing their need for a programmer's technical skills. This is particularly true of large companies.

The problems facing newly-minted programmers have been attracting attention in such publications as *Business Week* and the *New York Times*. "There is something of a glut at the junior end of the market," notes Ed Shaw, president of Setford-Shaw Associates, a New York employment agency that specializes in data processing jobs. However, Shaw adds, "data processing is still an open field. Once they get jobs, people advance rapidly."

Dr. Stuart Fink, director of data processing at New York University's school of continuing education, echoes Shaw's words. The bright side, as he sees it, is the emerging need for programmers in small- and mid-sized companies. These companies need programmers for mini- and microcomputers. Another ray of sunshine: Software houses that develop and sell programs represent a growth area for programming employment.

Getting a first job as a programmer may be a problem. If employment proves difficult for you, you might have to consider:

- getting your foot in the door at a company where programming positions exist by starting at a lower level—like computer operator, technician, secretary, word processing operator, receptionist, mail clerk, or messenger
- doing temporary work at companies that use computers to make a good impression for future consideration when programming openings develop
- volunteering—to get experience—at public corporations, nonprofit companies, religious organizations, community service groups, public schools
- pursuing advanced training with apprenticeship programs or some type of unpaid work experience
- using technical background as credentials for other entry-level positions like technical writer or customer support rep

Be aware that programming can be a dead-end first job. Or the position may not suit your interests and aptitudes. If this is the case, you must examine your background—training, experience, hobbies, likes, and dislikes—in terms of your preparation for other high-tech positions.

Identifying a suitable first job involves long range planning. Where might the job lead? What opportunities does the company offer in training and promotions? In many high-tech companies, your first position can set you on a career route that can prove difficult to alter later on. What would you like to be doing in five years? In ten? Do you want to move into management or remain in a technical area?

These questions may sound premature now. The fact is that a first job represents the initial step on a career ladder. It's possible to change direction or start over at a different company, or to make a lateral move, but it can be difficult. And it can mean time wasted.

Selecting your first job may mean rejecting a job offer or two. And that can prove difficult when time is passing and you're growing tired of searching for a job, your funds are running low, you're ready to work, and you have no guarantee that your ideal job exists. Turning down a job offer is a risk, but accepting the wrong job can be a serious mistake.

THE MYTH OF THE IDEAL JOB

You'll have to weigh any job offer against the likelihood of your landing the job of your dreams. How tight is the economy? Do you really have sufficient background for the job you'd like? If not, will the position you've been offered help you learn what you need to advance? Is the position likely to expand as you learn? Does your ideal job exist? Are you being realistic? Can you afford to wait?

THINKING AHEAD

To be specific, if you're trained as a programmer, consider whether you want to stay with programming, or move into systems analysis. If you're leaning toward a career in programming, ask yourself whether you want to handle an entire project from start to finish yourself or work on one part with other team members. What programming language do you want to work with? (Consider not only your interest, but the popularity and usage of the language.) What type of applications are you drawn to—business or scientific? Do you prefer working with mainframes, mini- or microcomputers? Do you prefer software houses, computer manufacturers, or the banking and insurance industries? Are you interested in consulting, financial analysis (both uncommon employers of inexperienced, freshly graduated job seekers), computer publishing (books, magazines, journals, newspapers, software), computer training (public and private colleges, universities, trade schools, secondary schools, elementary schools, company training and customer support divisions)? Do you want to be on the research, design, and manufacturing end—the vendor side—or on the end-user side?

Are you attracted to telecommunications, data base development or management, military applications? Use this book (particularly Chapters 1 and 4) to educate yourself about the diversity of high-tech areas. Highly-specialized areas like robotics, artificial intelligence, and genetic engineering will demand very specific advanced training. Computer graphics (including CAD/CAM, computer-aided design/computer-aided manufacturing) is a growing field that can offer excellent opportunities to those with the proper training. (Entry positions would be as junior members of the technical or sales staff.)

COMPUTER OPERATIONS

Let's take a closer look at specific entry-level opportunities available in the high-tech world. If you're technically inclined—or just interested in working directly with computers—and you've rejected programming as an option for the present, you might want to pursue computer operations. These positions pay less than programming, and mobility is generally more limited. However, an assertive individual might be able to advance from such positions as:

- computer operator or peripheral operator
- computer technician or field engineer
- data control clerk
- (tape) librarian

Keypunch operators and data entry positions are low-level jobs in predominantly female "ghettoes" that offer minimal opportunity for advancement. Improved technology is eliminating many of these positions as well. Avoid these jobs.

TECHNICAL WRITING

What entry-level positions might be suitable for you, if you rule out computer operations? Do you have experience working on a school newspaper, magazine, or yearbook? Have you taken technical writing or illustration classes? Do you have a portfolio of your work to demonstrate your skills? Do you have an interest in presenting information in a logical, clear, easy-to-understand way? Do you have a creative flair?

If you answer any of these questions positively, you might qualify for the following positions:

- technical writer
- documentation specialist
- advertising copy writer
- technical illustrator
- public relations writer
- editorial assistant
- computer or data processing publications journalist—staff or free-lance writer or editor—or member of business staff or production staff (proofreaders, paste-up artists, designers)

The number of computer-related publications has been growing at a phenomenal rate. Some use "stringers," free-lance reporters or writers who send in copy (writing) from major cities across the country. If you're interested in a high-tech writing career, one excellent way to break in —and build your credentials—might be as a free-lancer.

EDUCATION OR BUSINESS

If you've taken education courses or have teaching skills, you might be drawn to such positions as:

- customer support representative
- trainer
- computer literacy teacher
- teacher in trade school or community college (advanced degrees often not needed for these positions)

If your primary interest is business, and you view computer know-how as a tool, you might look for a first position as a management trainee in a wide variety of industries (banking, insurance, transportation, health services, wholesale and retail, utilities, manufacturing, construction, etc.). Specific coursework might qualify you for a position as:

- accountant
- marketing assistant
- financial analyst/trainee
- management consultant
- administrative assistant

SALES

If you've ever sold *anything,* are familiar with computer products, and know yourself to be persuasive and personable, computer sales may be the right first step for you. Keep in mind that high-tech sales will require familiarity with technology and an ability to demonstrate a product and compare it to the competition's. Opportunities exist in:

- computer retail stores
- computer manufacturing companies
- software houses
- computer peripheral manufacturers
- telecommunications companies
- computer graphics companies
- robotics companies

Your clients could be small, medium, or large businesses, individuals, schools, or libraries. What type of working environment do you prefer? Business or academic? Do you want to be at the same site daily or change location frequently? Would you be comfortable with your pay totally dependent on commissions?

WHAT EMPLOYERS LOOK FOR

Be prepared to hear these discouraging words repeatedly: You can't get a job without experience. In high-tech industries, this statement was less true than in other fields when computer science training was in short supply. It's a tighter market now and employers can afford to be more selective.

Employers look for experience and education, but they're also concerned with how you'll fit in as a member of a department and of a company. And surprisingly, many interviewers rate something higher than experience, education, and personality. In the words of Susan Jekiela who interviews applicants for sales positions at Hewlett-Packard in Rockville, Maryland, it's a "sincere interest in doing the job and doing it well. It's more important than any particular background. The bottom line is asking for the job, not waiting to be offered it."

When Maryann Haas, customer support center manager for Digital Electronics Corporation (DEC) in Boston, interviews job seekers for trainee positions, she looks for technical knowledge and maturity, initiative, and "something in their track records that tells me they're hard workers." That something may be a degree with honors or a record of employment while attending school. Whatever your something extra is, call attention to it!

PERSONAL PREPARATION

Perhaps the greatest adaptation involved in making the transition from student to successful job candidate is learning the meaning of professionalism. For professionalism—a catchall term for a blend of such diverse traits as confidence, modesty, maturity, understanding of how business is structured and how it works, company loyalty, and image—sets the serious job seeker apart from the student.

As a student, performance counts. How you dress, how you present yourself, whether you understand the politics of your school or not generally have little bearing on your grade. As a potential employee, all that changes. High-tech companies—though they may appear more informal and fluid than less innovative businesses—nevertheless have a working, if unspoken, code of professionalism.

Your letters, résumés, telephone calls, and performance at interviews must all reveal your professionalism. It's what will make you appear a likely member of the company team. Be professional in:

- your first contact with a professional employer
- what you include on your résumé and how you present the information
- how you dress
- how you interact with others
- how you speak (tone of voice, what you volunteer, how you respond to questions, what you ask)
- how you conclude interviews (with a handshake? by asking for the job?)
- how you follow up interviews (thank you note, phone calls)

We'll explore how professionalism relates to the major elements of your job hunt—résumé, interview, follow-up, dealing with an offer or rejection—as we go along. Here, in your first introduction to professionalism, let's concentrate on the professional attitude you'll need to embark on your journey to land a satisfactory first job.

As a female you may have been brought up to believe that tooting your own horn was "unladylike." Actually, evincing confidence about your abilities falls under the category of professionalism. A job candidate must be her own publicist. So before you begin to prepare a professional—and winning—résumé, ask yourself these questions:

- What work experience have you had? Include full-time and part-time jobs, summer positions, apprenticeships, internships, volunteer experiences, and school projects.
- What were your responsibilities in each position? Provide a job description for each job.
- What are your major accomplishments? Where have you shown initiative? Creativity? Commitment?
- What are your business skills? Identify them as technical, interpersonal, or personal. (Hint: Being a nice person doesn't count; getting along well with others does.)
- What courses have you taken that are particularly relevant to the positions you seek? List them.
- Who could provide you with the best references? A professor or teacher would be helpful, but employers are likely to desire a reference from someone who has supervised you in a work setting as well. Seek individuals who know you and your work well. Ask for their assessments before asking them to serve as references.
- What types of entry-level positions does your background suit you for?
- Which appeal to you? What are your interests? Values?
- How could you slant your experience when seeking different positions?
- Are you willing to relocate? Where would you consider living?
- When are you ready to begin work?
- Who can help you in your job hunt—not just with professional recommendations, but also in referrals, information, and support? Line them up.
- What salary do you need?
- What working conditions do you prefer?
- What career direction appeals to you? (What would you like to be doing in the future?)
- What entry-level positions would provide you with the necessary training and experience to follow your desired career path?

- What kind of people do you want to work with? People like yourself or different from you?
- How important is it to you to *like* the people you work with? To respect them? To be liked or respected?
- How complex a job do you want? Performing what and how many different functions?
- Which work environment would you prefer—profit or nonprofit, large or small organization, competitive or supportive atmosphere?
- How independently do you want to work? How closely do you want to be supervised?
- How important is challenge to you on the job? Comfort and familiarity?
- What personal satisfaction do you want to derive from your work? Are you looking to accomplish tasks, learn new things? Are you looking for another "family" at work? Or do you prefer anonymity—or a state somewhere in between?
- Do you want a nine-to-five job, or are you willing to take on a job that demands more time—to handle problems as they occur and to meet deadlines?
- How much stability do you require? How easily do you adapt to change?

RÉSUMÉS

Why Prepare a Résumé?

Before focusing on the do's and don't's of résumé preparation high-tech style, it's worth noting that many individuals secure jobs without ever showing a résumé. Some career counselors—a minority—argue that résumés are unnecessary. And for an individual with limited (or no) experience, such an argument might be tempting.

Robert Hochheiser, author of *Throw Away Your Résumé,* contends that a résumé is merely "a list of ingredients," and as such, résumés "are the worst way to sell your services." He advocates a more personalized strategy and recommends more pointed letters as a substitute.

In a letter, it's true, you might be able to give more weight to what you have to offer an employer, while you gloss over inexperience. However, most employers will—at some point during the consideration process—request to see your résumé. And not being able to comply with the request, whatever your explanation, may well brand you unprepared and suggest your lack of interest or business savvy. Besides, a carefully designed résumé can also "accentuate the positive"—in a more acceptable way.

A résumé represents your educational and professional credentials, highlights your skills, and announces your job or career objective. As such, it is to some extent comparable with any other applicant's résumé. At least this is the reasoning behind the résumé's acceptance as the premier tool for winnowing out unsatisfactory job candidates. Interviewers often believe that if a job candidate cannot handle the task of preparing a decent résumé (and that includes neatness, accuracy, legibility), she or he will be similarly sloppy or careless on the job. Since your résumé often precedes you, it's critical that it advertise your qualifications and spur employer interest.

What is a Résumé?

The word résumé, as you may have guessed, comes from the French. It means "summary." Bear in mind this definition as you prepare your résumé. It should be brief, concise, and informative. It should *not* include every course you ever took nor every task you performed at each of your jobs. Your résumé should be designed to arouse interest, not bury it.

It is imperative that you be honest on your résumé. Facts can be verified easily. An applicant who lies on a résumé will be judged untrustworthy. And trustworthiness is a trait highly rated by high-tech managers. If there is information about yourself that you consider unflattering—or detrimental to your chances for employment—omit mention instead of lying about it. Or face facts squarely and include an explanation. (For example, if you took six years to graduate from college instead of four, you might include your date of graduation only, or explain that you held a full-time job while attending college or graduated with a double major in computer science and technical writing.)

Some companies will ask you to fill out applications instead of submitting a résumé. Unless you are told specifically that your résumé can substitute for a completed application, fill the application out despite its apparent redundancy. You don't want to be judged arrogant, conceited, lazy, or ill-prepared.

Take note that a résumé mailed alone packs half the wallop of a résumé accompanied by a cover letter. A cover letter provides you with an opportunity to get personal, to interpret and elaborate on the data in your résumé, and to show something more. This might be knowledge of the company and its needs, your intense interest or enthusiasm, and your intentions (to call at a specific time to arrange an interview).

Types of Résumés

Now that we've placed the résumé in a proper perspective for the high-tech market, let's get down to basics. There are three kinds of résumés:

1. chronological (also called traditional)
2. functional
3. combination

While chronological résumés are most common, you as an entry-level job seeker would probably be better served with a functional or combination résumé.

A chronological résumé lists your professional experiences, beginning with the most current. If you lack high-tech education or experience or are reentering the work force after time out, or are changing your career direction, a chronological résumé would draw attention to your weaknesses.

A functional résumé lists the skills required for the positions you are seeking, and provides illustrations of your experience for each skill. If your education is on target (you've majored in computer science, electrical engineering, data processing, information systems, marketing, business, or mathematics—or you've taken many courses in these subjects—or you've completed a vocational school program), include brief descriptions of specific courses taken or knowledge acquired. For example, if you're applying for a programmer-trainee position, list computer languages you're fluent in, and name specific computers you've programmed. Be sure to use active, not passive, verbs—as in "mastered COBOL," not "was taught."

If you're applying for a position as a computer salesperson, you'll want to highlight your sales and technical abilities. If you've sold anything, emphasize your sales skills, *not* your inexperience in selling computers. In a functional résumé you might include phrases like "met" or "surpassed quota" or "chosen salesperson of the year." Look for transferable skills from your background and emphasize them. Did you sell high-ticket items for family use? Say so if your goal is to sell personal computers, video games, or educational software for home use.

If you have teaching experience and are looking for your first high-tech job, highlight your background in a functional résumé for positions as trainers, customer support representatives, or CAI (computer-aided-instruction) designers. Use verbs to indicate what and how you taught (active), not what your students learned (passive). Emphasize creative teaching methods you designed, number of students you taught, teaching awards won.

A combination chronological/functional résumé emphasizes functional skills but includes a work history. No matter what type of résumé you choose, two important questions arise:

1. Should you open your résumé with a job or career objective?
2. In what order should you place the major categories of infor-

mation contained in your résumé: education, employment history, personal information, and references?

Experts argue about the necessity of including a job or career objective in a résumé. When you are seeking your first high-tech job, specifying one desired position might be too limiting. Moreover, job titles vary tremendously (see Chapter 4) from organization to organization. You would do best to express your job objective with a general description of the position you seek, such as "a programming position that makes use of my fluency in COBOL, FORTRAN, and APL." Even this objective might prove overly specific if you're willing to learn other languages, or if you'd be open to a position as a trainer, customer or market support representative, or salesperson.

Despite these precautions, you *should* include a job objective. Make it as broad as possible, while being realistic about your qualifications. Or tailor your objective to the particular job you are applying for. Save your *career* objective for discussion at an interview.

Your job objective alerts a potential employer to the likelihood of your personal objectives matching the present needs of the company. It's true that some jobs are created for special individuals (more on that later in this chapter), but your résumé's job objective alone won't do that. And a career objective that looks five or more years into the future may be read as threatening ambition in a candidate for an entry-level position. Rather than set your sights on a particular future position in your job objective, indicate your (safer) desire to learn and grow. If that's too threatening to a potential employer, you're better off without that particular job.

Back to the second question: In what order should you place the major categories of information on your résumé? To answer a question with a question, which category is more important to you? If you have no formal education relating to computers—nor any related work experience—but you've created programs for yourself with your home computer, or you've been an active member of a user's group, or you've used word processing or financial spreadsheet software in coursework or at home, then the PERSONAL category is the one you should emphasize by leading off with it.

If, on the other hand, your personal information is unrelated to the job you are seeking, give it less prominence by placing it toward the close of your résumé. (Other interests can demonstrate that you're a well-rounded individual, so do include them.) Recognize that while aerobic dancing or films may be extremely important to you, they have little bearing on your qualifications for a high-tech job (generally).

If you're fresh out of school, EDUCATION will probably be the category you'll want to emphasize most. Be sure to list the names and addresses of schools attended, dates attended, degree(s) received and date(s) awarded, honors or awards if any, major (and minor, if any), and specific courses and grades if supportive of your qualifications. If you've done research under the supervision of professors well-known in your field, include their names and describe the project(s). Also describe independent work that illustrates special effort, innovation, or creativity. Include your class standing and grade point average if they're impressive.

And lastly, if your WORK EXPERIENCE illustrates your qualifications more fully than your education and training or personal experience, list that first. Begin with your most recent position. Include full- and part-time positions, summer jobs too. Provide dates of employment, names and addresses of employers (companies and immediate supervisors), your titles, your responsibilities, and your special on-the-job accomplishments. Mention any promotions or awards received from your employers or from industrial or professional associations.

REFERENCES are typically listed at the end of a résumé. Be sure to ask individuals who know your work and your character for their permission to be listed as references. Provide names, addresses, and telephone numbers for individuals or for your school placement office if it will maintain a reference file for you. If you prefer to direct potential employers to different reference sources, depending upon the position, you should write "References available upon request." Or you might consider following the example of one successful high-tech job applicant who clipped glowing recommendations from her professors and supervisors to her résumé. It caught the eye of Susan Jekiela, field marketing manager at Hewlett-Packard in Rockville, Maryland. The applicant's initiative and the quality of her references set her apart from the other candidates.

Other information that you deem pertinent should be listed under ADDITIONAL INFORMATION. This might include membership in professional societies or trade associations and leadership positions held, if any. You might also feature your membership in community organizations or volunteer experience that demonstrates skills relevant to business (such as operating or programming computers, publicity or advertising work, fund-raising, training or instructing, supervising or managing, research, budgeting, purchasing, scheduling, and shipping).

There is no need—and it serves no good purpose—to list information such as height, weight, color of eyes, hair, marital status, or number of children. Similarly, including a photograph should be taboo. This information has no bearing on your employability. It's not necessary to include your age or birth date either. And don't mention your previous or current salary—or the salary you're seeking. There is time enough to discuss this at an interview, or when a job offer is tendered. Including the salary desired before you're more familiar with a job's requirements and the market's going rate could lead you to sell yourself too low—or too high. Don't commit yourself to a desired salary before you have to. (Writing "SALARY: Negotiable" might be taken as a sign of excess malleability. It's best to avoid mention entirely.)

The length of your résumé depends on what you have to

include. As an entry-level job seeker chances are good that you can fit all the required information on a single sheet of paper. If you have important information that can't be squeezed onto one page, don't be constrained by the arbitrary rule that says résumés should not exceed one page. Remember that readability is crucial, and your cover letter can elaborate on any information contained in your résumé. Don't forget to list your name, address, and telephone number at the top of your résumé. If you can't be reached by phone, arrange for messages to be taken. If a potential employer can't reach you, the best of résumés counts for nothing.

Résumé Checklist

Here's a quick checklist to run against your résumé before you have it offset printed on quality paper:

1. Is your information attractively and neatly spaced on the page? (Never print on the back of your résumé.)
2. Is it free of typographical errors? Of misspellings?
3. Has a friend or relative proofread your résumé? Verified that information is clearly and accurately presented?
4. Have you included all pertinent information?
5. Have you double-checked all information?
6. Have you used active, not passive verbs?
7. Have you ordered categories of information in terms of importance, leading with the credentials you believe are most impressive?
8. Have you used underlining and capital letters effectively?
9. Have you been as concise as possible? Is all information included relevant to your employability?
10. Have you been specific about your skills, accomplishments, knowledge, and work responsibilities?
11. Have you avoided the repetition of personal pronouns?
12. Is your grammar correct? Sentence structure proper?

13. Does your job objective achieve a balance between specificity and generality?
14. Is your experience listed in reverse order, with most recent first?
15. Have you included personal traits that relate to the job you are seeking—like self-starter, energetic, enthusiastic, persistent, creative?
16. Can you think of a better way to present information?

(Don't be *too* innovative. You don't want any part of your résumé to turn off a potential employer.)

COVER LETTER

Once armed with a batch of shiny new résumés you'll need to prepare an eye-catching cover letter to introduce yourself to a potential high-tech employer and call attention to your attached résumé. The double-thrust of your résumé and cover letter should be to interest an employer in scheduling an interview with you.

Your cover letter shouldn't merely repeat or summarize the information contained in your résumé. Rather it should stress or elaborate reasons why you are particularly well-suited for employment with that particular company. Here is the place to show off your research into the company's products or services, means of operation, organizational hierarchy, new developments, etc. Knowledge garnered from reading the organization's annual report and newsletters or articles concerning the company or its managers or employees in trade, business, or general interest publications—or from conversations wtih company employees or those employed in the industry—demonstrated here proves your interest and enthusiasm. Relate your skills or background in a company-specific way and you've helped to answer the question "What can you do for us?"

Résumés are often skimmed for *facts,* cover letters read for *tone*. Cover letters offer you the opportunity to present yourself as a human being. A cover letter should be business-like, but allow your interest and enthusiasm to show. Support your case for taking an employer's time in an interview. Thank him or her for considering you. Be specific about following up your mailing with a telephone call to arrange an interview at your mutual convenience. Include a telephone number where you can be reached—and specify when. Be polite and friendly without being ingratiating.

Never begin a cover letter with the salutation "Dear Sir," "Dear Sir or Madam," or "Gentlemen," when you can locate a specific name instead. Unless you're responding to a blind box number advertisement (no individual or company name provided; address given only as a post office box number), this should be possible. Call the company. Ask for the personnel department or office manager, then pose your question. And be sure to spell the name correctly, using the title you are given—Ms., Mr., or Dr.

Keep your cover letter brief. Save some information for an interview. Three short paragraphs should be sufficient to introduce yourself, explain your purpose for writing, highlight your qualifications and background, and explain how hiring you would benefit the company.

Remember the purpose of your cover letter and résumé is *not* to secure you a job offer but to land you an interview. *Asking* for that interview requires that you be assertive.

Well-written cover letters are not redundant, even when an advertisement specifically requests that you respond only with your résumé. By sending a cover letter with your résumé you indicate your willingness to put forth special, individualized effort. Just as you tend to respond more positively to a personal letter than to a mass mailing, so does an employer.

Cover Letter Checklist

Before sending your cover letter and résumé, answer these questions:

1. Have you typed your address and the date on the right-hand side of the page at the top?
2. Have you included your addressee's title with her or his name and address?
3. Have you personalized the salutation with the correct name?
4. Have you mentioned how you learned about the job opening?
5. Have you identified the specific position you are seeking?
6. Have you briefly highlighted your pertinent educational and occupational experience?
7. Have you requested an interview at the employer's convenience?
8. Have you indicated where and how you can be reached? Included your telephone number?
9. Have you indicated your willingness to provide any additional information needed?
10. Have you promised to follow the letter with a phone call? Have you specified when you would do so?
11. Is the letter neatly spaced on the page with sufficient margins all around?
12. Have you and a friend or relative or career placement counselor proofread the letter? Checked grammar and spelling?
13. Have you balanced your tone between modesty and assertiveness?
14. Has your tone remained consistent? Have you remained aware of your audience throughout the letter?
15. Did you close the letter with "Sincerely," unless you have a personal relationship with the potential employer?
16. Have you signed your full name?
17. Have you indicated that you are enclosing your résumé?

INTERVIEWS

Once you've prepared strong résumé and cover letter prototypes (you'll want to individualize them for specific positions and companies), you'll need to prepare yourself for what is often the biggest stumbling block to a satisfactory job hunt: the interview.

The very idea may send butterflies fluttering in your stomach, perspiration moistening your brow, and prompt your hands to tremble. First piece of advice: Relax! It's normal to feel nervous about presenting yourself to strangers for consideration, but confidence breeds respect. Remind yourself of your qualifications. Remember that while a prospective employer is looking you over, you too should be evaluating the employer.

Not every job you interview for will be suitable for you. Yet it *is* worth arranging as many interviews as you can to perfect your interviewing technique, build your confidence, and increase the number of your contacts. It will also increase the likelihood that you'll discover the existence of other openings. Impressing a potential employer with your worth—even when a particular position appears at first glance inappropriate—can pay off with an expanded or altered position at the same company, or with an additional supporter and contact who may better your knowledge of *hidden* job opportunities elsewhere.

Don't lose sight of the fact that anyone you encounter on your job hunt can assist your efforts—or hinder them. Don't make enemies. Rudeness to a receptionist or secretary translates as an inability to get along with people. Similarly, your behavior during your job search will be viewed as predictive of your performance on the job. Tardiness for an interview may be taken as a sign that you don't really want the job or that you will perform it sloppily if hired. Inaccurate information—whether an outright lie or a slight exaggeration—will take you out of consideration if detected before a hiring decision has been made (What high-tech company can afford a dishonest employee?), or result in your dismissal if detected afterwards.

Here's a quick summary of what's critical to a successful interview:

- preparation
- confidence (knowledge of your value)
- honesty
- sensitivity
- poise
- modesty
- professionalism
- enthusiasm

INTERVIEW PREPARATION Let's explore these factors. Preparation for an interview involves familiarizing yourself with the questions you are likely to be asked, formulating answers in advance, researching the company, and preparing questions to ask the interviewer.

What questions are typically asked of a high-tech job applicant? Here's a lengthy list of possibilities. Formulating answers to difficult questions now—whether they're open-ended or highly specific—will minimize your nervousness during an interview. Try your hand at answering these:

- Tell me about yourself. (An invitation for a brief introduction, not a long-winded monologue.)
- What are your strengths?
- What are your weaknesses or limitations?
- What are your major accomplishments?
- What are your career goals? What would you like to be doing in five years? Ten?
- Where do you see yourself in this company in five years? In ten?
- Why do you want to work for this company?

- Why do you want this job?
- What do you know about our company?
- How could you contribute to the company?
- What are your qualifications for this particular job?
- Why did you choose to enter this field?
- Are you looking for a temporary or permanent position?
- What qualifications do you have that you feel will make you successful in the field? In the company? In the position?
- What have you done that illustrates initiative?
- Have you plans for continuing your education?
- Are you interested in relocating?
- What courses did you like most? Least?
- What percentage of your school expenses did you earn? How?
- Do you feel that you received a solid education?
- What are your salary requirements?
- Would you be happy working for us?
- Would you be happy in this job?
- Do you think you can handle this job? Why?
- Have you suffered any serious illnesses or injuries?
- What are your hobbies? Interests?
- What do you think determines an employee's promotability?
- Do you have a timetable for advancement?
- Do you like working independently? Do you like working as part of a team? Which is your preference?
- What job have you enjoyed the most? The least? Why?
- Whom can we contact for recommendations?
- Are you having other job interviews?
- What other jobs are you considering?
- What would be your ideal job?
- Do you prefer working in a large or small organization? Why?
- Do you know people employed at this company? In the industry? What have they led you to expect?
- How did you happen to apply here?
- Whom do you admire in the industry? Why?

- What are the advantages of work in this industry? Disadvantages?
- Can you work well under pressure? Can you meet deadlines? Are you available to work overtime?

It's not necessary to memorize your answers to these questions. Formulating answers in advance offers you the support of well-thought-out responses at the tip of your tongue. There's sure to be an unexpected question now and then, despite all your efforts. Take it in stride. Remember, admitting you don't know or you're not sure is legitimate, especially for an entry-level job seeker. And being open about your future has a great deal of validity in the fast-paced, ever changing high-tech world. (In five or ten years, jobs that don't exist now may pique your interest.) Admit your excitement—and flexibility—about changing technology. You want to appear decisive, but not permanently wedded to ideas that may prove outdated.

ILLEGAL QUESTIONS Certain interview questions are illegal, yet they are frequently asked of women nevertheless. Handling them can be a delicate matter. It is illegal in the United States (according to Title VII of the Civil Rights Act of 1964, as amended by the Equal Employment Opportunity Act of 1972) to ask about a job applicant's:

- marital status
- number or age of children
- future plans regarding pregnancy, childbearing, and childcare
- race
- religious or personal affiliations
- age
- height and weight

To ask these questions can run an employer the risk of a discrimination suit, and heavy fines if convicted. Smart interviewers don't ask illegal questions. However, many interviewers do, and their reasons may range from simple curiosity and friendliness to a personal or company bias. Slapping a potential employer with a discrimination suit every time one of the above questions is posed may be an overreaction—and one with far-reaching negative effects on your career development. On the other hand, meekly answering an illegal question is not the correct strategy either. Rather ask the interviewer for clarification. If you're still interested in the position and in the company, you might respond to the question "Are you married?" this way: "Does your company prefer married or single applicants?" A rare interviewer will fall into the trap of answering that question. Similarly, questions about children can be fielded the same way or with the dignified answer, "I believe that's personal information." Respond to these questions without sarcasm or anger and direct the conversation back to your qualifications for the position in particular and to your career in general. ("I want to assure you that I'm very committed to my career, and I'm very interested in this position in particular because . . .") If the questions can't be deflected in this manner, only you can decide whether you want to answer the question or remind your interviewer of the illegality of his or her question.

QUESTIONS FOR YOU While most job applicants—particularly those seeking their first jobs—prepare themselves to *answer* questions during an interview, they neglect to work up any questions of their own. And well-planned questions can go far to prove your intelligence, interest, enthusiasm, and preparation. Poorly planned questions can suggest that your priorities are misplaced or that you don't listen well, or that you're ill-prepared. Avoid these questions at a job interview:

- What are the benefits?
- How much vacation time would I have?
- How high a salary can you offer me?
- How soon would I be promoted?
- What does this company do or produce?

The last question reveals total ignorance and lack of interest in a company. If you do your research, you can ask more specific questions. Questions relating to salary and benefits should be saved until you are asked about salary or offered the job. Bringing up the subject prematurely—legitimate though these questions are—can brand you as insufficiently interested in the position itself. Instead, questions like the following should impress a potential employer:

- Is this a new position? If so, why has it been created?
- If not, can you tell me about the last person to hold this position? What was his or her background? How long did the individual stay in this position? Has the person been promoted? Been fired? Moved laterally? Resigned?
- Is this position commonly considered a springboard for other positions?
- How does this position—and the department in which it fits—relate to the company as a whole?
- Ideally, what type of person would you want to fill this position?
- Can you tell me who would be my immediate supervisor? (If that person is not your interviewer) Can I meet him or her?
- How large a department is it?
- How does the size and concentration of the company compare to that of a year ago? To the time of its inception?
- Is it the company's policy to promote from within?
- What do you consider the best aspects of the job? The worst?
- What would all of my responsibilities be?
- Is there a company management style? Can you describe it?
- What has the turnover rate been in this position?
- Am I being seriously considered?
- Is there anything else I can tell you or send you to add further support for my application?
- Do you know when you will make a hiring decision?
- Can I look forward to hearing from you, whatever your final decision is?

At an interview, let your interviewer take the initiative about how to greet you (with a handshake or without), when to sit, and when you should depart. Always allow sufficient time in your schedule for an interview to take place. Feeling rushed because of a tight schedule will damage your presentation. However, don't overstay your welcome. (Many interviewers stand to indicate the close of an interview, or physically see you to the door. A verbal signal like "I appreciated meeting you. Thank you for coming in," or "I hope to get back to you by the end of the week," serves the same purpose.)

The tone of an interview is largely determined by the interviewer. If you're lucky, the interviewer will be friendly and seek to maintain a relaxed atmosphere. This offers you the best opportunity to showcase your abilities. If you're unlucky, however, you may be in for a stress interview—or other variations such as a group interview, successive interviews, a negative interview, or an interview involving specific assessment tests (programming, personality, and polygraph).

STRESS INTERVIEWS The stress test interview, not uncommon in high-pressured, high-tech companies, is designed to assess how you respond to pressure. It's not a fun experience, but it may be an excellent guide to the conditions you'll actually face on the job. You may find an interviewer barking questions at you, rudely interrupting your answers, vehemently disagreeing with you, continually questioning your qualifications, and in general, coming across as an aggressive, abrasive individual. You may feel inadequate, cowed, or turned off. The question to ask yourself is, "Is the interviewer trying to alienate me from the company, show me what I can expect on the job, or see how tough I am?"

If a high-pressure work environment is not for you, it's important to know before you start. However, keep in mind that an interviewer's personal style may *not* mirror the company's. Always try to make the best impression you can to secure the job offer. Then you should consider fully what the working conditions would be like and whether it's the right offer for you. (More on how to assess a job offer later.)

GROUP INTERVIEWS Group interviews can take two forms. Either you are joined by other applicants to face questioning by one or more interviewers. Or you alone are interviewed by more than one interviewer simultaneously. Group interviews can be disconcerting when you've expected either more individual attention or less.

As an entry-level job seeker, it is doubtful that you'll face more than one interviewer at a time. Group interviews of this sort are generally reserved for more experienced job candidates. However, a company intent on filling many entry-level positions quickly may well decide that interviewing applicants in groups is the most expedient method. This can make it difficult for you to make yourself stand out in the crowd without being too overbearing. Make certain that all your qualifications are presented and discussed fully, ask all the questions you need to ask to understand the job as well as possible, and understand what degree of competition *and* cooperation are being sought—and demonstrate both.

A group interview can be helpful in that it reveals other interviewees' styles or presentations for you to learn from. Questions or comments from other applicants may also open your eyes to additional information about the company or the position. If you're faced wtih a group interview, stay calm but remember to assert yourself.

SUCCESSIVE INTERVIEWS As an eager seeker of your first job, you may feel euphoric after an interview where you know you made a good impression—and then feel dejected to discover that the interview you deemed all-important was only the first in a string of interviews that await you. The number of interviews you face for one job depends on the importance with which the particular position you are interviewing for is viewed by the company, the size of the company, the power of your potential immediate supervisor to make her or his own hiring decisions, the relative success or failure of the previous holder of the job you seek—and who hired him or her, how involved middle and upper management like to be in the hiring process, and how involved junior staff is allowed or encouraged to be in approving new staff.

Successive interviews typically begin with those who have the least power actually to hire you—but who do have the power to dismiss you from further consideration. Therefore you can never dismiss any interviewer as insignificant. It is critical to make an excellent impression every time, without showing bad grace about repeating your qualifications or dealing with an interviewer who may not understand the technical details of your background or the exact requirements of the position you are interviewing for.

If you start with an interview in the personnel department, be cognizant of the fact that your interviewer will probably be particularly skilled at representing the company to its best advantage and conducting an interview well (two key functions of personnel staff). That means you may well encounter less polished interview techniques as you progress in consideration, but you'll learn more about what you'd actually do, with whom, and what employment at the company would be like. Be prepared to learn from each interview—and keep your eye on the pot of gold at the end of the rainbow.

The bright side of receiving a job offer after a series of interviews is that you know that more than one company member is rooting for your success.

NEGATIVE INTERVIEWS The negative interview is a close cousin to the stress interview. As you might expect, the interviewer who practices this variety emphasizes the negative aspects of the position. Why? Some interviewers believe that this discourages all but the committed. In reality, the effect can be to discourage all applicants. If you encounter this approach, you'll have to work hard to achieve a balanced picture of the position in question. Rarely is a job all bad. Ask the interviewer about what you can learn. Find out the career progress of previous job holders. Explain why you believe the position might be a good first job for you, then ask for a response. ("Am I wrong?" "Do you disagree?")

Keep in mind that a negative interviewer may be correct: The job may not be for you. Listen closely and state your objections when they arise. ("I'm not afraid of hard work." "I understand there'll be occasional weekend work, and that won't be a problem.") If the interviewer spells out a job that has no redeeming value, don't descend to "Well, I need the money," or "I'll take it until I can get something better." These reasons never impress a potential employer. Convince the employer that you want the job because you want to work and believe you can learn from the position. Save more penetrating questions for after you receive the offer.

TEST INTERVIEWS Whether you're applying for a position as a programmer, technician, or sales representative, a potential employer might insist that you prove you're qualified by taking a test. It might take the form of a written test or a practical test, an examination of your technical skills, aptitudes, personality, or honesty. Can you refuse to take the test? Yes, but it can seriously hamper your candidacy.

If you fear performing badly on a test, you might ask how heavily it will be counted, or whether you can substitute a school transcript, a written essay, or recommendations. If not, you might as well give it a shot. It offers the employer another means of comparison, but it is rarely ranked as highly as your performance during an interview.

If you'd like to prepare for the test, you might practice taking programming tests available in book form from Arco, a publishing house, or study the personality profiles in Chapter 5 for the particular job you are seeking or the aptitudes required, as discussed in Chapter 4. A polygraph (lie detector) test is usually required only in positions where you'd be handling large sums of money or company secrets. These tests are not 100 percent accurate, but an employer may rely on the results nevertheless.

Some high-tech companies may request that you write a program or repair a computer or create a sales presentation as a practical test of your abilities. While this will require more time and effort than the usual hiring test—an interview alone—it can offer you the opportunity really to strut your stuff. By doing an excellent job you can prove your talent and abilities in a more immediate and forceful way than any verbal or written description of your abilities can.

INFORMATIONAL INTERVIEW This type of interview is one in which *you* set the tone. What's directly at stake here is not a job offer, but information gathering. Knowing how to arrange and profit from informational interviews sets apart the dedicated job seeker from the casual job seeker. And as with employment interviewing, there is an art to informational interviewing.

First, you must understand why informational interviews are important. How do you know which is the best career direction to pursue? Which high-tech companies put the most emphasis on promoting from within? Where is the industry headed? What qualifications are considered most critical to success in the field? Who are good role models? What are typical career paths? What is it like to work in the industry? At a particular company? In a specific department?

Reading this book answers many of these questions. Individuals you approach for informational interviews can fill in more details for you—and add to your network of contacts. While a job offer is not directly at stake, impressing those you seek information from can, further along the road, lead to a job offer. You can demonstrate that you're eager to learn, you know how to listen well, you're capable of systematic research and taking initiative, you know how to follow through, and you have energy and determination.

These are all highly valued traits in a high-tech employee. And an informational interview allows you to show these traits without the pressures of a job interview.

How to arrange informational interviews? Buttonhole "information sources''—politely and personably—at trade association or professional society meetings or conventions. Ask for names of people from friends, relatives, fellow students, teachers, and professors. Be on the lookout for those employed in high-tech fields at informal gatherings. Introduce yourself and request time to ask questions at a later date. Call people mentioned in articles in trade, business, and local publications. Try the authors of these articles. Visit or call employment counselors who specialize in placements in high-tech companies.

Many people you contact for information will be flattered. Others will suspect that your real motive is to ask for a job or for direct help in locating or landing one. Still others will be too busy to help you. Explain why you are contacting them specifically and indicate the kinds of questions you'd like to ask. Be flexible about scheduling a meeting. And show your appreciation.

At an informational interview, introduce yourself, your background, your interests, and your present career goals. Ask for specific information and request guidance. Have questions prepared in advance. Take notes as you listen to answers. Ask who else might be able to help you further. Follow up on every name you are given. And lastly, *don't* ask for a job. That would leave the impression that you arranged the meeting under false pretenses. You should ask if your informant knows of any openings in the field, and if she or he would keep your résumé on file should anyone inquire about an individual with your qualifications.

FOLLOW-UP

Whatever type of interview you experience, and whether you consider it a success or failure, *always* follow up with a letter of thanks. This letter should be polite and businesslike. After a job interview, address what has been discussed at the interview or the impression you believe you made. If you fear you made a negative impression, strive to correct it by focusing on why you feel you would be an asset to the company. Don't come on too strong or use the letter to beg for the job. Be dignified, repeat your interest, and sign off with a statement like, "I can be reached at . . .," or "I will call you Friday afternoon." Call at the promised time to thank the interviewer again. (Don't overdo it.) Express your continued interest and ask when a decision can be expected.

Many applicants fail to follow an interview with these simple courtesies. And when competition for a job is keen, this extra sign of interest and consideration can turn the tide in your favor.

LOCATING POTENTIAL EMPLOYERS

How to locate companies where you'd like to work? Here are some major sources:

- advertisements (want ads) in local newspapers and trade publications
- companies that schedule recruiting interviews at your school
- companies featured or mentioned in newspaper, magazine, and newsletter articles (follow up on companies reporting increased profits, new offices or plants, or those planning relocation)
- companies that manufacture products you're familiar with and find impressive (identify manufacturers, sales outlets, and advertising agencies if used)
- companies recommended to you by teachers or professors, school placement counselors, fellow students, individuals already employed in your field, members of trade associations and professional societies
- listings in such reference books as *Peterson's Annual Guide to Career and Employment for Engineers, Computer Scientists, and Physical Scientists; COMPJOB; Thomas' Register of American Manufacturers; Standard Directory of Advertising; Standard & Poor's Register of Corporations, Directors, &*

Executives; Dun & Bradstreet's Million Dollar Directory; Moody's Industrial Manual; F&S Index of Corporations and Industries; Encyclopedia of Associations; and *College Placement Annual*

- articles about high-tech industries and specific companies indexed in the *Computer Literature Index* and the *Reader's Guide to Periodical Literature*
- directories available from local Chambers of Commerce of area employers
- state industrial directories
- Yellow Pages of telephone books
- listings of companies with corporate memberships in industry groups
- lists of sponsors and attendees at computer-related and high-tech trade shows (attend to familiarize yourself with products, companies, services)
- companies who make appearances at job fairs and career days
- open houses sponsored by employee-hungry companies (generally held for experienced workers)
- computer user's groups (many members are employed by high-tech companies)
- job banks at trade groups, professional societies, or career support groups

IDENTIFYING JOB OPENINGS

In addition to identifying companies where you'd like to work, you'll also want to pinpoint actual openings. Eighty-five percent of jobs are unadvertised, so where do you find out about them?

- state employment offices
- referrals from contacts
- trade association and professional society job banks (like that of ASIS, American Society for Information Science, in

Washington, D.C., and the ACM—Association for Computing Machinery—Computer Science Employment Register at the Department of Computer Science, University of Pittsburgh)

- school placement office
- classified ads
- articles in trade and business journals, newspapers, newsletters, and tips from writers and editors
- management consultants in your field
- high-tech recruiters or employment agencies that don't require fees from job applicants (F/P or fee paid positions where the fee for the service is paid for by the employer)
- listings in such trade magazines as *Computerworld, Software News, Datamation, Byte, InfoWorld, Infosystems, Microcomputing, Interface Age, Dr. Dobb's Journal, Popular Computing, Personal Computing, Parity, The Computing Teacher,* and *Computers in Hospitals*
- networking
- classifieds in newspapers with national listings (like the *New York Times* and *Wall Street Journal*)
- federal government job listings
- mailings to potential employers presenting your qualifications
- advertising your qualifications and the job you're seeking in trade, business, and local publications

DEALING WITH REJECTIONS

If you're rejected by a high-tech employer, don't let it destroy your confidence. And don't give up! Try to learn *why* you were rejected, how to make yourself a stronger candidate. Could your résumé be more effective? Is your interviewing style too passive or too aggressive? Is your background insufficient for the job you were seeking? Have you selected the best references? Ask the interviewer for feedback to help you in future interviews. Show that you're interested in improving your candidacy. Listen carefully to what you are told. Request elaboration if necessary, without putting the employer on the spot or insisting on your suitability.

If you discover that you failed to convince the employer of particular qualifications, you'll know what to emphasize next time. If the employer doubted your real interest in the job, you'll have to work on demonstrating your interest in the future. Were you well-prepared for the interview? Were you poised and confident? Did you focus on what you could do for the company, not what the company could do for you? Did you demonstrate more interest in salary and benefits than in the job itself? Did you come across as excessively—or insufficiently—ambitious? Was your desire for further education seen as threatening? Did you appear unprofessional in dress or manner?

Don't turn a post-rejection discussion into a third-degree interrogation. Thank the interviewer for her or his time and helpful feedback. And act on that advice as you continue your job hunt.

Remember there may well be external factors that influenced an employer's decision not to offer you a job:

- the background and experience of previous job holder(s)
- the other applicants; their credentials and their contacts
- time pressure—how long the position has been open, how urgently the position must be filled, who is overworked as a result
- who must agree to hire; who has the power to make the final decision
- resources—how much they can pay
- the existence of training programs, formal or informal, within the company
- the size of the company
- the interviewer's own biases and mood
- the state of the economy
- the company's financial situation
- the opportunities available for promotion in the company

- security of interviewer's position
- the interviewer's evaluation of the risks incurred by hiring you
- the interviewer's subjective evaluation of your compatibility with other company employees
- the company's unspoken criteria for the job
- the interviewer's unspoken criteria for the job
- the company's "personality"
- company politics
- your contacts or references—and interviewer's biases concerning them
- job requirements you have not been informed of
- sex discrimination
- age discrimination
- a negative report from a previous employer or reference

These are factors that you may not know about—and thus cannot counter. Accept that rejections are not always understandable. Swallow your disappointment and maintain a professional relationship with the interviewer. Remember that any contact you make during your job search may prove helpful in providing you with other leads and recommendations and in helping you achieve your goal.

JOB OFFERS

After the often exhausting and occasionally disheartening process of hunting for your first high-tech job, you may be ready to seize your first offer gratefully, as a well-earned reward for your efforts and the long-awaited end of your journey. It's not so simple. Never accept *any* job offer on the spot unless you have given careful consideration to all the consequences. Request time to consider the offer and assurance that the offer will remain yours while you consider it. Accepting the wrong first job doesn't doom your career forever, but it can lead you in the wrong direction. And while you don't want to stay with your first job forever, remaining for a reasonable time will indicate your stability—and save you from the rigors of engaging in another job search immediately.

How to evaluate an entry-level job offer in a high-tech industry? Ask:

- Why is the company hiring at this level?
- What is the company's reputation as an employer? Does it exploit or reward workers?
- Does it offer on-the-job training or promote continuing education? (Arrange seminars and workshops for employees? Provide tuition remission? Value education?)
- What is the turnover rate at the company? In your position? In your department? If high, why?
- Is it company policy to promote from within?
- What is the company's financial condition? Status in the marketplace?
- What are company goals? Are they realistic?
- Will the particular position offer you good experience?
- Can you handle the job as it's presently structured?
- What are office politics like at the company?
- What is the company's management style? Do you find it congenial? Tolerable?
- What is the company's attitude toward entry-level workers? Toward women?
- Is the salary offered within the appropriate range for the position? Can you negotiate a higher salary or better benefits?
- Can you work well with those who would be your co-workers? Your managers?
- Is there the right blend of independence and supervision for you?
- How large is the company? Will you be highly visible or anonymous? Does that suit you?
- Can you learn from the job?

- Can the position lead you in your desired career direction?
- Is the position the best possible first job for you?
- If you are compromising, is it a reasonable compromise? One you can live with?
- Can you afford not to take the job?
- Is there an opportunity to grow in the position? In the company?
- Can you make good contacts on the job?
- Can you make friends on the job? Is that important to you?
- Are you first choice for the position?
- Do you have any negotiating clout?

Recognize that no job is perfect. Compromises may be called for regarding salary, scheduled reviews, responsibilities, and location. Weigh the importance of getting your foot in the door against the likelihood of "stubbing your toe" in the process, or entering through the wrong door, or being closed out. Do you think you can do better? Is your belief realistic?

Turning down a job offer—without another in your pocket—involves risk. There is no guarantee that something better awaits you. Draw up a list of pros and cons. In the final analysis, the decision is yours. A good, first high-tech job exists for you. Is it your present offer? Make your decision and then get on with your career—and your life!

Moving and Advancing

Landing your first high-tech job is only the first step along your career path. Odds are better than even that at some point in your career you'll be interested in changing jobs—advancing into management, moving up the technical career ladder, making a lateral move into a position you find more interesting or suitable, or even leaving to start your own company.

You must decide what your professional goals are. In the fast-paced high-tech world, technology is continually transforming the jobs that already exist while creating new jobs. Futurists predict that lifetime education will become the norm as people adapt to a work world in a constant state of flux. It is impossible to predict accurately one's every career move, but your attention and planning are necessary for a career that *you* control. Lack of attention and planning can leave you adrift in a series of less than satisfactory jobs or stuck in a dead-end position that stymies your professional growth.

Ask yourself these questions:

- How far would you like to progress professionally?
- What personal sacrifices are you prepared to make to further your career?
- Do you require immediate gratification?
- How much stress are you willing—or able—to tolerate?
- What are your financial requirements, short-term and long-term?

- What status and occupational titles do you aspire to?
- Are you more interested in furthering your technical knowledge or sharpening and using your management skills?
- What are your priorities in your personal and professional lives?
- Do you have management ability?
- Are you sufficiently competitive and assertive (or can you learn how to be) to force your way into management, if necessary?
- Do you have the proper temperament for positions up the ladder and for advancing your career? How ambitious are you?
- Can you strive for and accept the visibility that is a prerequisite for advancement?
- What is your natural leadership style? Is it respected and successful in your company or in your industry?
- Are there opportunities to advance from the position you now hold?
- Do you have decision-making, problem-solving, and risk-taking abilities?
- Can you visualize the whole picture at your company? Can you understand the nature and interrelationship of your goals, your department's goals, the objectives of other departments, and the broader goals of the company as a whole?

UPWARD MOBILITY

Upward mobility has been at the heart of the American dream for a long time, but the realities of today's economic and social conditions have put brakes on the career aspirations of many. The coming of age of the Baby Boom generation (eighteen- to thirty-five-year-olds) translates into increased competition at all levels of employment, from first job to higher rungs of the career ladder. And the expanding use of computers in business to gather and analyze information has resulted in a shrinking number of middle managers. Economic conditions too have had the effect of paring the ranks of management, flattening the corporate "pyramid" or hierarchy.

A 1983 Special Report on management in *Business Week* (April 25, 1983) reported the following changes that would-be managers should be apprised of:

- Modification in corporate structure: The increasing usage of computers allows executives to gather data for decision-making directly. The end result is greater productivity. Middle managers who once gathered, monitored, edited, and "second-guessed" this information are redundant. Many staff managers are being fired. In a 1983 study by Professor Eugene E. Jennings of Michigan State University, an estimated one-third of the 100 largest industrial companies were cutting management ranks.
- Middle managers who hold on to their jobs must become generalists, not specialists.
- Fewer MBAs are being hired. Those who are hired will stay in one position longer and find lateral moves more likely than true promotions.
- Computer, marketing, and manufacturing skills are becoming more important than the formerly highly valued skills of analysis and gamesmanship. "Line" managers will become more necessary—and have greater job security—than "staff" managers. (For definitions, see page 184.)

As gloomy as the picture for middle managers appears, aspiring senior managers—especially women—also face less than rosy futures. Dr. Shoshana Zuboff, a social psychologist who studies the effects of technology on the work force, believes that the numbers of senior managers will also decline.

These predictions indicate the difficulties you may face in your pursuit of career advancement. Add to these the highest level of unemployment among managers and administrators in non-farm industries since World War II—as reported by the Bureau of Labor Statistics in 1983—and the picture is far from encouraging.

But now for the good news. And there *is* good news. Those women with computer, business, interpersonal, and communication skills are at a decided advantage on the path upward to high-tech management. And as you've seen in Chapter 2, women as a group tend to have an advantage over men as managers.

Before highlighting the high-tech positions that offer the most upward mobility, let's get personal. What does it take to be a high-tech manager? Do you have the goods? Do you have the desire? Take this quiz to find out.

Are You Management Material?

1. Are you willing to match your goals to those of your company?
2. Do you have greater loyalty to your company than to your department?
3. Are you comfortable speaking to those on your level as well as those above and below?
4. Do you have a good sense of how your company is organized—who has power, who reports to whom? What the objectives, constraints, and problems of different positions and of different departments are?
5. Are you willing to keep up with industry developments by keeping abreast of trade publications, books, and special reports? By attending meetings, trade shows, and exhibitions? By joining and participating in professional associations?
6. Do you expect continuing education to play a large role in your professional life? Are you prepared to take advantage of in-house training programs, as well as workshops, seminars, classes, and programs outside of work that can teach you new skills or sharpen those you need to get ahead? Are you prepared to pursue an advanced degree, like an MBA, to be more competitive?
7. Do you like being visible in your company? Do you know how to be more visible, to call attention to your

accomplishments, and to maintain a professional image?

8. Do you have expertise in your company and industry that gives you confidence, respect, and power?
9. Can you be consistent and relate openly to others?
10. Do you see yourself clearly and realistically? Do you look to yourself for final approval?
11. Can you offer constructive criticism? Can you accept it?
12. Can you accept help and support? Can you provide it?
13. Can you make decisions quickly? Can you take risks?
14. Are you self-directed?
15. Are you skilled at making valuable contacts—inside and outside your company—for information and advice?
16. Do you understand how to trade favors for mutual benefit?
17. Can you be tough, firm, and assertive, without being aggressive or abrasive?
18. Are you self-confident?
19. Do your skills at persuasion and negotiation rate highly?
20. Are you well-organized and systematic?
21. Can you manage your time well?
22. Are you adept at defining priorities? Handling many projects and details simultaneously?
23. Could you schedule not only your time but that of your subordinates?
24. Is meeting deadlines more important to you than perfection?
25. Can you give feedback and ask questions when necessary?
26. Do you listen well to others?
27. Do you believe respect is more important than popularity?
28. Are you well-liked? Do you get along well with others?
29. Can you accept the possibility that attribution of credit may not always be fair?
30. Are you willing to be a trailblazer if there are no women managers in your company as yet?

That's it. Pencils down. How do you measure up? Score five points for each yes answer to questions 1 to 8, and one point for each yes answer on questions 9 to 20. Yes answers on questions 21 through 30 earn three points each. Negative answers don't score points.

Key

27 points or less: If most of your yes answers were to questions 1 to 8, you may understand business, but not have the right personality to become a manager. If most of your yes answers were for questions 9 to 30, you may understand the personal makeup of the managerial role, but not how business is structured.

28 to 54 points: You seem to understand the internal workings of business and a manager's place in the company hierarchy. Yet you may still have a distance to travel toward preparing yourself for the difficulties and compromises inherent in management.

55 to 82 points: Give yourself a pat on the back! You're management material. You understand the realities of high-tech management and are prepared to take your place in its ranks.

An aptitude and a desire for management, an understanding of the internal structure of a company, and self-knowledge are among the prerequisites for high-tech advancement. Don't despair if you scored below fifty-five points. Reading this chapter will help you raise your score. Let's examine which starting positions offer the best growth opportunities to build on these prerequisites.

TECHNICAL VS. MANAGERIAL CAREER PATHS

Depending on your training and educational background, your first high-tech job may be as a programmer, engineer, technical writer, salesperson, customer support person or trainer, or systems engineer. In high-tech companies an employee generally faces two—theoretically—parallel routes for advancement: technical and managerial. It's important here to make a conscious decision about your career direction based on your talents and goals.

If you're uncertain about which direction offers you greater professional growth, you should take note of the conclusions of a revealing study by Laurie Michael-Roth, a researcher at the Center for Research in Career Development at Columbia's Graduate School of Business. She found that despite the lip service paid to the notion of separate-but-equal career paths, the management ladder offered significantly greater prestige, power, and money. In fact, the technical ladder often becomes a "major 'dumping ground' for managerial rejects," rather than a respected route for technical talent.

In theory, technical ladders are designed to offer the same potential in salary and decision-making power as managerial routes. However, as Michael-Roth points out, this is rarely the case. Those with good technical skills—especially when these skills are in demand—may find management prefers to keep them performing technical functions, to the detriment of their professional growth.

If your preference is for the technical career ladder, examine your company's commitment to advancement and remuneration for technical staff. Your career might follow this route: entry-level trainee to programmer to programmer/analyst or instructor to systems analyst to specialist to consultant. Or a programmer might advance to senior and then chief programmer or project leader, like Diane Hajicek at MicroPro, a San Rafael, California software house. An electrical engineer might start as a technical staff member and advance to project leader or consultant, like Sandra Tuck at the management consulting firm, Booz, Allen in Rockville, Maryland.

An instructor might move into sales. A successful sales representative might be promoted to field marketing manager, like Susan Jekiela at Hewlett-Packard in Rockville, Maryland. After three years selling business computers at HP, she was promoted to her present position. Jekiela is responsible for recruiting and training all new salespeople for the Baltimore/Washington areas, lead generation (scouting new sales opportunities), marketing (shows and seminars), and developing new training programs for sales staff.

How to advance up the technical ladder?

- Keep up with the latest technical developments. Read trade publications, attend trade shows and meetings, participate in trade associations, attend workshops and seminars. Your sharpened technical expertise is your ticket upward.
- Make your ambitions known. Lobby for promotions. If management assumes you are content with your present position, you are likely to be overlooked for a higher one. "I had to make it known I was interested in the position," says Maryann Haas of her promotion to customer support center manager at Digital Electronics Corporation in Boston. "No one tapped me on the shoulder."
- Take courses or pursue an advanced degree or certification to add to your credentials. Sandra Tuck has taken graduate courses in telecommunications to help her with her job as technical consultant.
- While learning your job completely, groom your successor and learn about the next job up the ladder. Tackle tough assignments and claim credit for your achievements.
- Look for a mentor or sponsor, if possible. (More on this later in this chapter.) Or learn from a role model's example.

- Network with others in your field to make connections, learn about new developments and about job openings.
- Demonstrate that you can work successfully with others in teams, as well as independently. Prove your ability to accomplish tasks, meet deadlines, and motivate other team members.
- Always maintain a professional image and attitude.

ROUTES TO SENIOR MANAGEMENT

If your goal is to enter into the top ranks of management, you will have to make crucial decisions along the way about:

- line vs. staff experience
- relocation—gathering "multiple site experience," "bench" (technical) and field (away from the home office) experience
- undertaking a variety of assignments in a variety of departments to learn about the organization as a whole
- how long to stay in one company
- further education
- your personal life

Promotions that offer additional responsibilities and attractive salary increases may not always assist you in reaching your high-level goal. In fact, many promotions can change your direction and provide you with the wrong kind of experience, necessitating backtracking at a future date—to pick up needed experience—at a sacrifice in salary or job level.

A 1982 study of women in senior management at large American corporations undertaken for Korn/Ferry International pointed out some interesting differences in the profile of the average senior woman manager and that of her male counterpart:

- Only 21 percent of the women had relocated geographically.
- The average woman manager had changed jobs only three times during her career and had worked for her present employer for thirteen years.

- 30 percent of the women held graduate degrees, compared to 43 percent of the men.
- Only 40.7 percent of the women were married, compared to more than 94 percent of the men. Seventeen percent of the women had been divorced vs. 2.4 percent of the men. (And most women who had been divorced reported that their careers were a factor.) 27.6 percent of the women had never married, compared to less than 1 percent of their male counterparts.
- Women identified the routes to the top as marketing/sales, financial/accounting, and professional/technical. Few women held top positions in manufacturing, a critical "line" department.

Traditionally, line experience and the ability to relocate have been among the prerequisites for advancement into senior management. And women have occupied more staff positions and have been (or have been *assumed* to be) less willing to relocate. The precise definitions can vary from industry to industry, and from company to company, but here is a working definition of staff versus line positions: Line jobs and departments are those that contribute directly to the bottom line; staff positions do not and therefore are more vulnerable during recessions and company or industry shakedowns. In a company that manufactures personal computers, line departments would be operations, sales and marketing, and research and development. At a software house, programmers would be in line positions. At a bank or insurance company, programmers and other data processing staff would hold staff positions. In a consulting company, consultants would hold line positions.

Experience in staff positions—like technical writers or documentation specialists (unless the company prides itself on its documentation and considers it a significant selling point or its only or most important product) will count less toward advancement to the higher levels of a company. Similarly, experience in an organization's data processing department may only lead to rewards like director or manager of the data processing center or division, or vice-president of data processing or MIS (management of information systems)—but no further unless supplemented by experience in line divisions.

SALES/MARKETING ROUTE

The most popular route for women to high-tech management has been through sales and marketing. As the number of competitive high-tech products has risen—in personal and business computers and software and in advanced telecommunications products and services, to name just a few areas—sales and marketing have become increasingly important. Having a good product, as many high-tech companies have learned, is far from sufficient. It must be wisely and imaginatively marketed and sold aggressively.

Janice Miller, president of Women in Information Processing (WIP), headquartered in Washington, D.C., notes that the "people at the top are marketeers." It's a rare woman who has progressed far from a more technical path, such as through engineering, she adds. Many others agree:

Nadine Solomon, operations supervisor at United Technology Communications, Inc. in New York, says: "In general, sales still seems the easiest route to management." There are, however, paths into management through technical lines as well, she affirms.

Pam Jacobson, regional sales manager for videotex at American Bell in New York, believes that sales "gives you a broader view" than other departments. She thinks sales offers more possibilities for advancement because "technical people get locked into particular positions when they do what they do well." In addition, as a member of a male-dominated sales force—Jacobson's experience, and a common one at high-tech companies—a woman is highly visible. And visibility can lead to further advancement opportunities.

If you don't have a technical background, but you're a "quick study" and relate well to people, sales may well be your best route up the high-tech career ladder. You should research a product thoroughly to assure yourself that it's competitive and that the company is sound—well-financed and well-managed. A strong sales record can lead to further advancement, but it will mean turning your skills from making presentations and persuading customers or clients to hiring, training, coaching, and managing a staff. As a salesperson, your salary may depend totally on commissions for products sold. As a manager, your salary will be tied to a different kind of performance.

DATA PROCESSING/MIS CAREER ROUTE

If you begin your career ladder in the data processing department or MIS division of a corporation whose central product is not computer-related, you will have to step outside these divisions into departments more closely linked to the organization's core business. Otherwise your journey will end at the highest levels of your department, not of the corporation itself. To move up you will need to capitalize on managerial strengths—polishing your management skills while moving away from a narrow or overly technical business outlook. You will need to keep up with general technology but relegate the details of hardware and software to subordinates.

You should recognize as well that positions of responsibility in MIS/DP management are exceptionally high-risk and high visibility positions in most companies. Speaking from personal experience, Georgette Psaris, formerly a vice-president at Osborne Computer Corporation (now defunct), warns that these are dangerous positions to undertake. If serious problems develop, you can lose your job. Psaris took on the responsibility of developing a central computer system at Osborne and counts herself lucky to have found an excellent system. Her success augmented her reputation in the company and helped her advance further.

A survey conducted by Fred Held, a vice-president of Mattel

Toys, identified some common characteristics of executives who rose up through MIS/DP ranks. These abilities and experiences are ones you should develop if this is your planned route:

- a comprehensive understanding of the organization's operations, goals, problems, and personnel and management philosophy
- experience with EEO (Equal Employment Opportunity), Affirmative Action, data privacy, business ethics, OSHA (Occupational Safety & Health Administration) standards, union requirements, mandating retirement policies, environmental concerns, and consumerism
- demonstrated ability to manage a line function and perform a high-level staff position
- experience in a multi-location or multi-national company and understanding of its communications and logistics problems
- experience in developing and implementing short-, mid-, and long-range planning; ability to deal with top management during planning and to integrate DP plans with an organization's other functions
- ability to assign priorities, allocate funds, understand budget concepts
- ability to conform to set budgets, schedules, and end-product specifications
- success as a member of a middle management team; reputation as a productive and cooperative member
- experience using advanced technology to solve company problems and benefit an entire organization

FORMAL TRAINING FOR ADVANCEMENT

What kind of formal training will stand you in good stead for high-level advancement? An MBA may not be necessary, but courses in finance, business ethics, time management, government regulations, human relations, public speaking, effective business writing, and corporate politics would help. Lore Harp, former president and CEO of Vector Graphic in California, built up her company as she learned on the job about business practices. It was only after her company was well established that she decided to pursue an MBA at Pepperdine University. Her reasoning? She knew how she handled problems at her company. The academic program introduced her to other ways of doing things.

COMPANY SIZE AND ADVANCEMENT

Career paths differ at large and small companies. At smaller companies career paths tend to be more informal and the experience required more generalized. Responsibilities handled by two or more individuals at a large company may be combined in one position at a small company. Titles may signify less at small companies where responsibilities count more. Smaller companies may not have clearly defined career paths. More will depend on company success and growth. It may be expected that once you acquire sufficient experience, you will need to change companies to advance. On the other hand, a small company may provide the best learning opportunity and the faster route to advancement because movement may not be as rigid and more attention may be paid to the individual. And since a small company depends more on the service of generalists than on specialists, it may offer better training for management. You should be aware, however, that larger companies often have more status and thus may look down on small-company experience.

As a woman you should take note that EEO and Affirmative Action programs tend to be more prevalent at larger companies. Sex discrimination may be less common as a result. The climate at small companies may be a reflection of the founder's or president's or CEO's personal feelings and attitudes. This can work in your favor if that person is progressive and unbiased, but presents serious problems should the individual be chauvinistic and a seeker of employees in his or her image. While formal affirmative action programs may not be necessary to your advancement, their presence does indicate a sensitivity well placed in a still male-dominated industry.

What else might a large company offer over a small one?

- more in-house training, education programs, and formal tuition reimbursement programs
- formalized performance evaluations and scheduled salary reviews
- more opportunity to move into other departments or transfer to other locations
- more opportunity to make connections and network
- greater opportunity to find a role model, mentor and/or sponsor and to learn from others in your company

COMPETITION

High-tech organizations, like most others, are structured like pyramids—either with the middle ranks of middle managers shrinking (as discussed earlier) or swelling while the top of the pyramid shrinks (according to tech psychologist Shoshana Zuboff, Ph.D.). Whichever the case, competition is in order. Organizations are not structured to provide promotional slots for every junior staff member.

To many women, competition implies cut-throat tactics, dishonesty, secrecy—in short, a web of vice that appears ethically wrong and likely to ensnare the ill-prepared competitor in a shadowy underworld. Leery of competing with office friends or arousing negative feelings (such as resentment or disapproval) in others and ambivalent about winning and losing, many women hold stubbornly to the misguided notion that value wins out. They wait to receive their just rewards—and wait and wait.

A healthy attitude toward competition involves the recognition that competition can be engaged in with a clear conscience. However, you should be alert to the use of dirty tricks on the part of less-than-scrupulous competitors who attempt to take credit for your ideas or your work, minimize or criticize your efforts, privately or publicly, sabotage your efforts, attack you personally, or use personal or confidential information against you.

The best strategy for combating these dirty tricks is to meet your detractors head on, inform them that you know about their actions, and demand that they cease their misrepresentations. Often puncturing the screen of secrecy is sufficient to deflate dirty tricksters. If this doesn't work, don't attempt to fight back at their level—unless you have the skills and the stomach to handle a rough fight. Instead, try to rise above gossip and slander and prove yourself by working hard, calling attention to your accomplishments, making your own allies, and proving your commitment to your work.

How to compete fairly and cleanly?

- Promote your own performance, without denigrating that of others.
- Take on additional responsibilities, showing initiative.
- Always do the best job you can do.
- Ask for feedback and follow good advice.

COMPETITION VS. TEAM EFFORT

Being a successful and cooperative team member is necessary for most every advancement-oriented high-tech employee. In high-tech companies, the project team is often the unit of task accomplishment. A project team member must interact successfully with other team members, with management, support staff, and frequently with clients, customers, or end-users as well. Therefore, the high-tech employee must recognize when to subordinate competition to the greater good of the team's goals—

and to the company's goals—as well as the customer or client's needs. Failure to recognize these unchanging priorities can send you back to square one.

It's true that in some high-tech environments (particularly software houses), mavericks are the ones who succeed up to a point. For technical whiz kids, crackerjack salespeople, and super-imaginative programmers, being a good team member may be of little import. These individuals—because of the high level of their contribution to the company—can get by without the interpersonal skills required by less extraordinary employees. The maverick may have an abrasive personality or a completely independent manner. As long as her or his accomplishments are superior, rewards—like raises and promotions—will follow. However, entering the managerial ranks—where interpersonal skills count—will often be closed. And subversion by those antagonized by the maverick can often lead to her or his ouster. The best advice: Be a cooperative, friendly team member when necessary. It's worth the effort.

VISIBILITY

If you were raised to believe you should be seen and not heard—as many women were—you may have a difficult time overcoming your childhood training. While boastfulness and aggression are still frowned on at the office—especially for women—letting others usurp credit for work done well or for your ideas is downright foolhardy.

In a large company, not calling attention to your talents and accomplishments can render you invisible. And when it comes to promotions, one must be visible to be considered. In a smaller company, visibility is less of a problem, but self-promotion is still important.

How can you mount a successful self-promotion campaign? First, you need to affirm that ambition is healthy. It *is,* and that drive to advance makes visibility critical. As a member of the

Baby Boom generation, you'll need to compete for advancement at every step of your career. It's a matter of numbers. Make yourself stand above the crowd by:

- acquiring a reputation for good work
- being a hard worker
- being organized and efficient
- having strong technical skills
- communicating eloquently and effectively
- being likable and easy to get along with
- being willing to tackle tough assignments
- working well both in teams and independently
- being creative
- being able to take constructive criticism and act on it
- having a professional attitude and showing your company loyalty
- being able to follow company guidelines, formal and informal
- being a problem-solver

These abilities will make you an excellent candidate for promotion into management.

Visibility for attributes like those described above will further your career. The wrong kind of visibility will hamper your career progress. Here's a list of don'ts for the would-be manager:

- Don't dress flamboyantly. Dress-for-success cookie-cutter blue suits won't be necessary, but you should take clues from how managers at your company dress. If junior staff members wear jeans, but managers dress more formally, follow their lead. Dress up. And don't dress provocatively at work. You won't be taken seriously as a businessperson. Despite all the media attention given to dress-for-success ideology, it's no surprise that clothes alone do not make for success. However, earning a reputation for being well-dressed and well-groomed doesn't hurt. It's part of maintaining a professional image. (Note: Overdressing at a casual company can be an error. Harmonize with company style.)

• Don't bring your personal life into work. While it is a good idea to present yourself as a well-rounded individual with interests—and a life—outside of the office, don't confuse your priorities. At work, your personal life should take a back seat. You want to be recognized there for your professional accomplishments, not for your hobbies or family life.

• Be sensitive, but not oversensitive. Women professionals have often been accused of being too emotional and sensitive. Sensitivity is critical to the success of a manager, but oversensitivity earns you the reputation of being prickly and difficult to deal with.

While the office environment is becoming more humanized as more women—and younger, progressive men—make their way upward, there are still those who will be made uncomfortable by an "excessive show of emotions." This means that tears should be avoided.

Here are some tips to promote your professional visibility:

• Use memos to highlight your accomplishments; express modesty and share credit where it's due.

• Support your boss and discuss your career goals with her or him. If your boss is unsupportive, pursue contacts at higher levels.

• Join professional societies or trade associations. Be an active member by running for office, chairing panels, leading workshops, participating in seminars and conferences.

• Write for trade or business publications. Bylined articles or signed letters to the editor can help familiarize others with your name.

• Present your ideas in written form, and follow them up with in-person reminders.

• Speak up at company meetings. Be an active participant.

• Make your ambitions known.

FIRST-LINE MANAGEMENT

Moving into a supervisory role—entering the first line of management—may well be the most difficult transition for a working woman. Before you enter management (and supervisory positions are often not considered *true* management positions, though they are usually the first step into management), you are part of a team. Being on equal footing with your co-workers can offer a satisfying sense of camaraderie, a feeling of belonging to a group. A promotion changes both your relationship with your former co-workers and managers. In the case of former co-workers, this change is often for the worse.

You may face envy and resentment. Friendships may fail the test of your success. As a new manager, you must learn to cope with feelings of being apart. And you must learn how to exercise authority and delegate tasks. Helping and teaching those you manage are often the gratifying parts of the role. Assigning and scheduling tasks, checking quality, evaluating and criticizing performance, and assuming heavier responsibilities are generally the more difficult managerial functions.

There are three possible routes into your first managerial position. You may be promoted from the ranks; you may be brought in from a different department; or you may land your first supervisory position by moving to a different company. Which route is preferable?

There are costs and benefits to all approaches. If you are promoted from the ranks, your former co-workers may feel unhappy about their own lack of progress and hold your promotion against you. They may see your promotion as favoritism and resent it, you, and upper management. On the other hand, your former co-workers may be pleased at your success and eager to help you advance the department's work and improve working conditions. Much will depend on who was in your position before —or whether it was created for you—and how your predecessor or the new organizational structure is perceived. In addition, the circumstances of your predecessor's departure will play a part in how your promotion is perceived.

EVALUATING AN OFFERED PROMOTION

An offered promotion into management is a career opportunity for success *or* failure. Evaluate the offer by thoroughly investigating the situation you'd be entering. Are you being set up for failure? Will subordinates and managers be supportive? Will you be expected to follow your predecessor's record to the letter (an impossible task), or be allowed the autonomy to institute merited changes?

Whom will you be supervising? Individuals who competed for your job and still feel the sting of defeat? Or less-experienced employees who consider your promotion a learning opportunity for them and/or a burst of fresh air in a stuffy and stagnant department?

Is management offering you the opportunity to be a token or a trailblazer? Or to take on a simpler role, that of doing a job that needs to be done? A token will be proffered a title, but no real responsibility. A trailblazer will be offered a real opportunity, but one within a special context.

Many company onlookers will be rooting for a trailblazer's failure, just as others will be pinning high hopes on the trailblazer's success. For a trailblazer, the intensity of these hopes is high and the pressure great because much is riding on the individual's performance. If there have been no women in management at your company, you may be invited to take on the role of trailblazer and be first. The opportunity can offer tremendous satisfactions along with the pressures.

You should be aware that high-tech businesses, because they are younger than many others, often present greater opportunities for women in management. Why? Before companies grow and become more established, they tend to be more liberal in their policies and less formal in their career paths. Small, growing companies demand more of their employees at the outset, but the rewards—in promotions and financial rewards (stocks, royalties, bonuses, perks) if the company succeeds—are correspondingly higher.

The dark side of high-tech industries, however, is that many of them are male-dominated. A "young boy" (or "old boy") network may be in effect, unconsciously serving to exclude women. Only women with superior technical skills may be admired and "let into the club." Women without may be at a disadvantage.

The criteria for advancement vary from company to company. They may include a technical degree, a business degree, technical experience, managerial skills, interpersonal skills, connections, or some combination of these factors. Research the background and qualifications of those in management at your company or at a company where you'd like to advance. Do you have the appropriate credentials? Are you willing to pursue them if you don't?

TECHNICAL VS. BUSINESS SKILLS

Traditionally, the path to the top of a high-tech company has depended on technical expertise. The engineer or programmer who creates or improves the money-making product or service—or its means of production—generally has been the one who advanced. This is not always smart business. Technical pros are not necessarily good managers. And as this truism has been borne out repeatedly in high-tech company after company, new criteria for managerial positions have been instituted.

Yet technical expertise continues to play a large part in the selection of managers because it remains a highly respected credential. Building on your technical skills and experience can help make you a more attractive candidate for promotion. But if it's senior management you aspire to, concentrate on refining your managerial skills as well.

High-tech managers are well served by understanding the "bits and bytes" of their companies, but of greater importance is their ability to hire, train, and motivate those beneath them on the company ladder, and to add to the company's bottom line. Accomplishing tasks becomes less valuable than motivating others to accomplish tasks. A manager overly concerned with overseeing each and every detail of a project misses the larger picture. Such a person—often excessively tied to a technical outlook—will be ineffective as a manager.

The major problem for the would-be high-tech manager is that there is rarely a clearly defined career path to the highest level of management, no such labelled entry-level position, nor even middle-level positions. Instead, individuals tend to cross over from advanced technical positions into management. There are, it's true, managerial positions of lower rank in departments such as personnel or advertising, but these are rarely the best stepping stones for progress into the upper reaches of a company.

POWER

Power can derive from your position, expertise, information, responsibilities, and connections. Seizing power, holding it, and augmenting it are necessary political skills for professional advancement. Power has been defined by business experts as the ability to get things done. And acquiring a reputation for getting things done is a must for potential managers. Attempting to not play the power game can leave you out of the running for significant advancement.

Since lower positions are not loaded with power (or status), early on in your career you must concentrate on adding responsibilities, acquiring information, honing your expertise, and making connections to broaden your power base. While adding responsibilities carries with it the risk of failure, the risk you take is worth it because it establishes you as management material.

Take on responsibilities you feel confident you can handle. Be sure to draw attention to your increased responsibilities.

Maintaining a work diary, a description of responsibilities and accomplishments, is one excellent method of tracking your job development. A diary is not only a helpful reminder to yourself of your own initiative and professional growth, but also a valuable tool in negotiating future promotions that will legitimize the power you have acquired in this way.

In the high-tech world, power is often closely aligned with technical expertise. Even if technical skills are not required for a particular position, having them is often recognized as having earned your stripes. A technical degree or comparable experience can earn respect among peers, subordinates, and those higher up, which in turn can add to your power.

Too often women mistake the negative face of power for the *only* face of power. Power, they say, is a matter of dominating others and they want no part of it. They see power as a brute force wielded by insensitive autocrats who *force* others to accomplish tasks and meet goals. Too many women drop out of the competition for career advancement because of their misguided notion of power politics in the workplace.

To be sure, some managers do operate as autocrats. However, they are rarely *effective* managers. Effective managers *inspire* their staff, and recognize that power must be shared. They encourage and persuade staff, educate and coach them.

Power arises from self-confidence. If you know your position inside and out, you have power. If you know your position better than any of your predecessors or counterparts, you will have more power. And if you add to your position by assuming responsibilities outside your original job description, you not only become a more valuable employee, you add to your power as well.

You may have difficulty locating power in entry-level or lowly positions. Power derives from performing a function that needs to be done. Don't underestimate the importance of the functions you are responsible for accomplishing. And power derives from the expertise you utilize—and enhance—in the course of performing your job.

As a programmer, you could increase your power by studying other languages you envision becoming important to your company. As a sales rep, you can expand your territory or the number of products you sell, or add to your technical knowledge about the particular products you sell.

When your knowledge surpasses what is normally required by your position, you have enriched your power as an expert. This is particularly true if you can become the most knowledgeable person in your company about a specific area. When other employees come to you with questions or to seek advice, you benefit the company and your position in it.

Georgette Psaris's phenomenal career led her from a "lowly" marketing assistant position at Osborne & Associates, a consulting and publishing company, to a vice-presidency at Osborne Computer Corporation in five years. The key to her advancement has been her initiative in undertaking new responsibilities and adding to her expert power.

"In a small company you never want for additional things to do, additional responsibilities," Psaris observes. When she was hired in 1978, there was no international marketing of Osborne & Associates' computer books. Within three weeks she set up contracts in "about twelve languages" and an "exhaustive translation program." She lined up distributors and established a strong international sales department.

In less than a year, Psaris added to her position by undertaking responsibility for worldwide marketing and sales. A few months later she took charge of customer services, shipping, MIS, and the central computer system. She became the company expert on these functions.

The power of connections can also lead to advancement, so

effort made in making connections is time well-spent. This may sound cold and calculating, but forming relationships with the right people is critical. Who are these people?

- those who are successful
- those whose work can affect yours
- those who are in a position to evaluate your work
- those who could become mentors or sponsors

You shouldn't ignore less powerful people. Earning the dislike of co-workers or junior staff members can set you up with a host of detractors ready to sabotage your climb upwards. Of course, you can't please all of the people all of the time (some clichés are true), but do your best.

This advice bears emphasis because too often women feel sincerity is the coin of the realm in company politics. They don't understand why they should be politic about personal relations at work, such as putting effort into being found likable by those they couldn't care less about.

Sincerity *is* an honorable trait, but it doesn't always pay off. I'm not advocating dishonesty, but tact and good sense. Don't be insincere in your efforts at connecting. Be smart.

PROFESSIONAL IMAGE AND ATTITUDE

If you agree readily that factors like connections, expertise, experience, and training are important in professional advancement, you may be inclined to dismiss the equal weight given to professional image and attitude. To do so would be to commit a serious oversight. Maintaining a professional image in the high-tech world is crucial.

Professional image is more than a matter of what you wear, how your hair is styled, and whether you use makeup. It's how you present yourself—the total image you project—as you represent your company. If management doesn't think you have it, the odds on your being promoted into their ranks are very poor.

What contributes to professional image in the high-tech world?

- ability to communicate—to speak, listen, and respond effectively
- capacity to accept criticism, graciously if not gratefully
- ability to take initiative
- good grooming
- assertiveness without aggressiveness
- appropriate behavior (no flirtatiousness or being "cute")
- proper priorities (private life private)
- harmonizing with the unofficial (or official) company image (as represented by powerful or successful managers)

Mimicking the professional style of your boss or your boss's boss illustrates a lack of imagination. Besides success doesn't demand such slavish imitation. Rather, as an upwardly mobile professional, you should take lessons from your superiors at work. What are their communication styles? How fashionably or conservatively do they dress? How independent are they? How assertive? Keep in mind that if your boss is male or much older or much younger, what works for him or her might not be effective for you. In addition, a style appropriate for a manager may not be appropriate for you at a lower level.

Professional attitude, an integral component of your professional image, is a hazier area for examination. The definition may vary widely from one high-tech company to the next. In general, the attributes that form a professional attitude would include seriousness tempered by a sense of humor, collegiality, maturity, and trust blended with competitiveness.

If you're in doubt regarding what constitutes a professional attitude, analyze those of respected managers at your company. Focus on how these managers interact with others at all levels.

How friendly or reserved are they? When are they serious? When do they crack jokes? (Note that taking on the persona of a stand-up comic may work well in sales presentations, but fail abysmally at department meetings. And if you constantly forget punchlines or your humor runs to the ethnic or pornographic, step carefully or avoid humor entirely.)

How much trust do successful managers display in their subordinates? (Keeping a finger in every pie and refusing to delegate any responsibility represent unprofessional attitudes tenable only in the lower ranks of management.) How competitive are they with those on their professional level? Are they admired for their ambition or disliked for their cut-throat tactics, unkindness, or self-centeredness?

PERSONAL STYLE

The line between personal style and professional style is often a thin one. Some call personal style personality, others likability, and still others personal chemistry or social skills. If you've got it—that difficult-to-define something extra that makes people warm to you—use it in your career. It's your personal style which can provide an added boost up the ladder.

What makes for a successful personal style in the high-tech world?

- tact
- concern for others
- articulateness
- sincerity
- being a good listener
- friendliness
- warmth
- humor

A good personal style is not always synonymous with an effective managerial style. Examine what works in your company.

Being liked is nice, but being respected is more significant. If the personal style outlined above strikes higher-ups in your company as too "soft," you'll have to adopt a tougher personal style to impress them with your promotability—or move to a more congenial company.

SIGNIFICANT OTHERS

Role Models

Role models are individuals who have made their way up the career ladder and provide an example for you to emulate. Having a role model of the same sex can help convince you that you, as a woman, can aim high and succeed as well. You needn't have personal contact with someone you select as a role model. You may even take as your role model one of the successful women profiled in Chapter 1 or Chapter 4. A role model offers proof that the career development you desire is possible as well as illustration of how, when and why career development in your industry occurs—guidelines for you to follow (not slavishly) or a mirror to hold up to your own progress.

If there are no role models in your company, look to members of trade and professional associations or to women you read about in trade, professional, and business journals, newspapers, or books. Research the professional backgrounds of women who have succeeded with high-tech careers.

What were their first jobs? How much time did it take them to progress to a managerial position? How much time did they spend in individual jobs before moving up—or across? Did they spend their entire careers at one company or company hop? Have they assumed leadership positions in professional societies? Are they well-known outside of their companies? Outside of their communities? Outside of their industries? Do they give speeches, write papers or articles, or serve on panels or commissions? What are their academic credentials? Are they assertive?

Competitive? Have they had mentors or sponsors? (More on these roles later.) Have they been active networkers? Are they well-liked by their staffs? Is their leadership style authoritarian or democratic? Learn from their examples.

Mentors

Much has been written about the benefits of securing a mentor in a corporate career. In a high-tech company, the importance of a mentor is very much dependent on the size of a company. In a small company, an individual's performance and capabilities will be much more visible than in a larger company. A mentor would not be necessary. At a large company, however, a mentor could foster your professional growth by passing on news of unpublicized openings, praising your talents and performance to those higher up, informing you of office politics and both formal and informal company structure, and advising you on everything from dress to behavior to recommended career moves. Your response to this list of goodies may well be, "Sounds terrific, where do I get one?" A good question.

Asking someone to be your mentor is akin to a child asking another child to be her friend. The request rarely clinches the deal and often puts into words what should be suggested in a less forthright manner. It's more a matter of impressing a potential mentor with your professional worth and ability to rise through the ranks. After all, a mentor rarely needs your help. If you do well, it can add to her or his reputation, but it's only reflected glory. You're the one who stands to reap big benefits from a mentor's sponsorship. And that puts the ball in your court. How to attract a mentor?

• Excel at your work. Your performance must be high for a potential mentor to notice you. You must prove yourself extremely competent at the work you do, even though you've set your sights on higher ground. Performing sloppily on your present job, far from signalling your candidacy for bigger and better things, could take you out of the running. The issue is a bit murkier than this statement makes it appear, however. If you are presently employed as a programmer, but your goal is to become a manager, your technical skills can bring you notice. However, you'll need to indicate leadership ability and interpersonal skills to be considered a good candidate for promotion into management.

• Voice your immediate ambition. Performing well at your present job may not garner a potential mentor's attention if you don't express those ambitions which a mentor could help you attain. However, articulating *high* ambitions could sound brash, overly aggressive, or even threatening. Follow the middle path by presenting yourself as ambitious and realistic. Talk about your goals one step at a time. Too often women wait for their competence to be rewarded. There are more women stuck in programming and low-level supervisory positions than in middle and upper management. Their silence may be mistaken for satisfaction with the status quo. Would-be mentors who assume that you are contented with your lot will shy from offering you assistance to advance. If you are considered a valuable employee in your present slot, you must impress your bosses with the fact that you will be even *more* valuable if promoted. Don't be afraid to ask for promotions. A mentor will want to take on a good worker and an ambitious one.

• Promote your visibility. In a large company, those who might become your mentors may not know who you are. As one of many junior level employees, you must call yourself to their attention. How? Make contacts in other departments. Join trade and professional groups and networks where you can interact with possible mentors whom you could not come into contact with at work. Impress them with your worth and ambition. Send company-wide memos, write for your company newsletter or trade magazines. Call attention to community work—like your involvement in career counseling at local schools. Give talks at your local library or community center about your work or industry, or lead classes or workshops. Be friendly to possible mentors in hallways, elevators, and the company cafeteria or gym. Introduce yourself. Discuss your ideas concerning particular projects. Don't turn informal meetings into hard-sell sales presentations. Rather take advantage of casual interaction to make a good impression quickly. You should be aware that such efforts at self-promotions may not be viewed kindly by all. Some will consider you immodest, or worse, pushy. Many who believe that a job well-done is its own reward will be disturbed by your active lobbying for advancement. But self-effacement does not represent the path to management.

How will you recognize a mentor? Rarely does a mentor make a formal announcement. Rather the relationship might be initiated by a mentor praising your work or asking about your career goals. A mentor might be comfortable giving you guidance or putting in a good word for you with those in a position to promote you.

If you are lucky enough to attract an offer of professional help from a mentor, be careful to examine the offer for strings attached. Would the relationship be solely professional? Is help being offered because of your achievements or for nonprofessional reasons? If you are certain that your would-be mentor's motives are honorable, ask yourself how the relationship would be viewed by co-workers. Would they respond with resentment and jealousy? Would their negative feelings eat away at the possible pluses of having a powerful ally? Focus on your potential mentor's place in the company structure, both formal and informal. Does she or he occupy a secure position in the company? Is the individual well-respected? Has she or he helped others to advance? What has been expected in return?

Understand that allying yourself with a mentor means that a mentor can help you rise—*or fall*—with her or his own star. Hitching a ride up the ladder with a mentor could be dangerous business if he or she doesn't see your career as you do, loses his or her job, leaves to start his or her own company, and doesn't ask you to join, or does, but you don't want to, or leaves for another company and your company loyalty is questioned.

Betty Niimi, general manager of silicon chip producer Intel's Austin, Texas, operation, credits a mentor with assisting her career development. She met her mentor when they both worked at Univac. When her mentor moved to Intel, he offered Niimi a position there. She took it, but cautions others, "You always need to make sure that you're earning your own way, as opposed to relying on a particular sponsor." Today Niimi's sponsor is no longer at Intel. (A sponsor is roughly the same as a mentor. More on the distinctions in a minute.)

If you're prepared to take the risks of having a mentor or sponsor—you've balanced the possible costs and benefits, and the benefits come out ahead—recognize that a mentor will expect loyalty. You shouldn't abdicate responsibility for your career development and dump your aspirations into your mentor's lap. Nor should you always accept the word of your mentor. Instead, ask for advice about your career direction, and request explanation if your mentor's guidance conflicts with your own beliefs. If you decide to go against your mentor's advice, express gratitude anyway and explain why you are following a different course.

Sponsors

I've discussed mentors so far as if they represent the only type of professional support you might receive from others. In fact, professor of management and author Natasha Josefowitz distinguishes between mentors and sponsors. Mentors, she claims, can be individuals at your level or your immediate supervisor. Mentors have personal contact with you and advise you on specific steps to take in the present. But as helpful as mentors are, sponsors can go even further.

Josefowitz identifies a sponsor as an individual who has greater power in the company than a mentor. A sponsor is more highly placed but probably does not have regular contact with you. However, a sponsor is concerned with the future and will speak about your talents or performance to others. A sponsor may also introduce you to others on her or his level or above. By seeing that you get recognition for your achievements in the company, a sponsor adds to your company visibility while promoting you for consideration for higher positions in the firm.

A sponsor may also keep you in mind should she or he leave to found another company, join another firm, or be asked to recommend an individual to fill an opening at another company. A sponsor is a "coach" in your corner. Yet you may never know you have a sponsor—or who she or he is. To capitalize on the possibility, however, the same guidelines for seeking the attention of a mentor apply.

So don't overlook the importance of friends in high places—rarely can they hurt your career. But remember, while a mentor or sponsor might ease your path upward, not securing one does not doom your career to failure. Many successful high-tech women have made their way unassisted. You can too.

NETWORKING

Networking extends to you another option for maximizing your professional growth. And profiting from networks tends to be easier than securing a mentor. A network may be a formal or informal group, open to all or highly selective in its membership. The goal of professional networks is to further the careers of its members. The modus operandi? Contact-making for support and mutual advancement.

Women in high-tech fields have often felt isolated. Their companies may be male-dominated, and as members of a female minority, they may find advancement more difficult. Networks function as forums for professionals to discuss career problems and solutions and to disseminate news about job openings and opportunities.

Many women's networks present workshops and seminars on an ongoing basis to educate members about important management techniques like effective writing, listening, and speaking skills, and assertiveness training, as well as continuing technical education concerning new products and techniques. This knowledge is power, added credentials for the upwardly bound high-tech worker.

Formal job banks, job listings, and career counseling programs are also common. To name just a few associations with such services: The Association for Systems Management, Association of Computer Programmers and Analysts, Association for Women in Science, Society for Women Engineers. For addresses and a more extensive list of associations, societies, and networks, see Appendix I.

Formal job banks are only a small part of the story. It's a well-known fact that a staggeringly high percentage of jobs come through personal contacts. The U.S. Bureau of Labor Statistics has put the figure at 48 percent. Other labor experts estimate that the figure is far higher for management level jobs.

To progress into management, you need to know where and when there are opportunities. Network membership multiplies your contacts and the likelihood that you'll be in the right place at the right time to hear about those opportunities. In casual conversations at network meetings and conventions, you'll meet people in your field who are in a position to exchange news about openings and provide inside information about those openings. Members can and do refer other members, and extend the lessons of their own experience.

During the course of Betty Niimi's career she received a call from a headhunter about a position at Univac as director of sales and applications for minicomputers. How did the headhunter get her name? From the president of the New York-based Women in Engineering society, whom Niimi knew personally. "Opportunities do come up through the various organizations," Niimi says. "These types of contacts are probably the best type of leads."

The idea for women's professional networks grew out of the consciousness raising groups of the early 1970s. The emphasis of these earlier networks has shifted to professional concerns. As women watched men climb up the career ladder with the help of other men—the so-called old boys' network—they decided to institute the same kind of network for their own sex.

Becoming well-known for your participation in a professional network can be valuable for your career because your reputation in the group can serve as legitimization. A manager of human-computer interaction engineering at a 15,000-employee company (who prefers to remain anonymous) claims that her corporation came to her with a management offer after she had spent three years on the technical staff because of her visibility *outside* of the company in the professional community. She has moderated different professional society conferences and been a member of different society advising boards and commissions.

Before you join any network, be aware that participation requires time and effort. If you are a member in name only, you won't reap many rewards from the connection. On the other hand, attending meetings and planning speeches, discussions, and conventions takes time. As Betty Niimi puts it, "If you're going to be involved, you have to be able to contribute at a certain level."

The importance of having a support network behind you depends on many factors, such as where you live, your present position, the contacts you have already, whether you have a mentor or sponsor, the size of your company, how established your company or industry is, your personality type, and the particular network itself.

Betty Niimi believes that cities like Philadelphia and Minneapolis are less liberal about women managers than West Coast cities. Therefore, a woman with management aspirations in a more conservative locale would need a network's support more. Similarly, if you live in a small town, belonging to a regional or local branch of a national network can open your eyes to more managerial opportunities than may be available where you live now.

TIMING

When are you ready to move into management? Well, you may feel ready after a year on the job as a programmer or salesperson or trainer, but successful women in the field advise you to be more patient. Betty Niimi believes that five years is a reasonable amount of time to become a manager. It's sufficient time to acquire a "technical reputation and respect in the organization," she says.

"Moving into management after a couple of years is probably difficult and probably would be a detriment long term. Because if you don't have a reasonable technical base—if you get into management too quickly—it's like building a house on not too strong concrete. You need some technical understanding for management," says Niimi.

This doesn't mean that one should stay in one position for five years. Niimi advises not staying in the same job for more than two or three years.

"You psychologically sit down and say, 'Where have I come in the last year, what have I accomplished, what more can I learn from the job?' I do that every year."

In young, growing companies, it's reasonable to expect your advancement to correspond with the company's growth. And large companies, by the nature of their organizational structure, tend to have training programs—formal or informal—that promote movement for qualified employees—or possible lateral movement for the interested worker. (It should be noted, however, that companies in shaky financial condition—due to national or local economic problems or company- or industry-specific problems—will slow advancement if not pare staff outright.) If you are employed by a small company which is not on the move, you may find opportunity more limited.

Timing your professional moves is necessary to assure that you have sufficient training and experience—or initiative and inner strength—to handle a promotion and to save yourself wasted months or years in a no-reward or dead-end position. In the best of all possible worlds, you will be offered only the promotions you are ready for, exactly when you are ready for them. Reality, however, is more capricious. When an opening arises, your qualifications may be of less import than the need to fill a position quickly.

How to analyze the suitability of a job change? Here are ten things to do and questions to ask:

1. List your business and technical skills and grade your understanding and/or performance of each.
2. Match your skills to that of your predecessor. Are yours superior, inferior, or about equal?
3. What can you learn from the job? Does it represent a challenge? Will it equip you with the knowledge and experience you require for future advancement? Does it further your progress toward your career goals or redirect your career?
4. Is the opportunity more than a new title?
5. Will support be forthcoming from staff? From management?
6. Can you handle a difficult political situation if interference and static are likely?
7. Is the position offered with merited salary and all due responsibilities?
8. Is position "performable" or overloaded with responsibilities?
9. Should you suggest dividing the position? Are you being set up for failure—or for exploitation?
10. Would you be happy in the new job?

To repeat this chapter's theme: Careers *can* be managed and directed, although luck and happenstance do play considerable parts. Everything can't be controlled, but you can formulate a career plan replete with goals and intermediate steps, and work toward implementing it to the best of your ability. You'll have to be flexible, adapt your plan to changing conditions and new trends in your company, your industry at large, and your personal growth.

Be prepared to monitor your progress periodically, but avoid an overly rigid timetable. Don't be concerned if your initial plan proves unrealistic. Adapt to evolving conditions and be alert to new opportunities.

Patience: Is it ever a virtue in the high-tech world? Is job-hopping or company-switching frowned upon? The answers to these questions are yes and no. Practicing patience may be wise or foolhardy, depending on the particular circumstances. If you've been in one position for more than two or three years, are you being fair to your career?

Ask yourself:

- Am I happy in my present situation? Do I enjoy my work? Am I still learning and growing?
- Do I have sufficient experience and the necessary skills to handle the next position up the ladder?
- Am I still in the same position because no opportunity for advancement presently exists in the company, but one should be opening up soon? Or will I never be seen as promotable in this company? Am I being realistic about the opportunities available to me here?
- Would competing companies offer me greater opportunities? Or would I have to put in my time there before being able to reach my goal?
- Do I need further training or education to advance?
- Do I need to speak of my ambitions in a louder, more forceful voice—and to more powerful people? Am I being overlooked because I have not made myself visible?
- Does the blame lie with me or with the company, or is it shared? Am I being honestly apprised of the facts or lied to? Am I being exploited or groomed to advance? Are the company's and my timetables in step?

Only you can decide whether you are being wisely patient or

foolishly passive. Analyze the situation calmly; then act. And bear in mind the words of Amy Wohl, president of Advanced Office Concepts, Inc., a Pennsylvania office automation consultancy and publishing house: "Women are often not aggressive enough. They're too polite. You have to ask for what you want if you want to get ahead."

TIME MANAGEMENT

A trait related to a good sense of timing is a superior time-management ability. The worker who is offered a promotion is not necessarily the individual who works fifteen hours a day, seven days a week, but the person who accomplishes as much in less than half the time. Management wants to see results, not show. Making constructive and efficient use of a normal workday indicates one's self-mastery, a prerequisite for managing others. How do you rate at effective time management? Do you have the:

- ability to set priorities
- ability to divide projects into component tasks and accomplish them one by one
- ability to begin work without procrastinating
- understanding that perfection is rarely possible
- sense not to take on the impossible; the sense to know when to say no and when to request assistance
- the sense to take breaks when necessary to renew energy and combat boredom

FLEXIBILITY

Flexibility counts for a great deal in career planning because all the variables can never be figured precisely. In high-tech fields, new products and services or new developments in production make dramatic entrances—and exits. They can open up exciting new career vistas or permanently block hitherto upwardly mobile career routes.

Look at Pam Jacobson's career in the promising field of videotex. Jacobson began her career as a junior-high-school art teacher after earning a B.A. in art, French, and history in 1973. She saw the limits of a teaching career and abandoned it after a year to enter telecommunications. In the next nine years she rose up the ranks at AT&T, surpassing her sales quotas, adding responsibilities, and earning promotions.

When AT&T developed its first commercial videotex product, Jacobson saw the wave of the future in the new technology and jumped on the bandwagon. Videotex had not been invented when Jacobson first entered the field. Her willingness to be flexible about her career development has enabled her to take advantage of new technology.

The benefits of flexibility are also borne out by the early career of Susan Jekiela, now field marketing manager at Hewlett-Packard. After graduating from college with a degree in mathematics, she taught junior-high math for two years in small towns. She decided that she wanted more financial recognition as well as the social life that a city could offer. Though she liked teaching and working with children, Jekiela decided to pursue another people-related career. She landed her first high-tech job as a sales rep at Burroughs, selling custom-designed computer products.

"When I first started, I was scared to death," she remembers. She overcame her fears and was flexible about relocating and entering an industry entirely new to her.

"I think sales is similar to teaching—you're using the same people skills," Jekiela says. She's discovered that she prefers working with professionals to working with junior-high children. "It's much more interesting to see how business works," she adds.

For Bonnie Barit, also at Hewlett-Packard, flexibility came into play when a friend called her while she was a systems analyst at Honeywell. The subject of the conversation: a position at

Hewlett-Packard. The move involved switching from mainlines to minicomputers. It was a good opportunity, but Barit was initially leery. She decided to be flexible and moved to HP as a staff systems engineer. After three-and-a-half years, she is now a systems engineering district manager.

In addition to flexibility concerning career direction, you'll need flexibility to deal with problems, setbacks, and confusion. Candidates for promotion must be able to cope with—and overcome—the problems and pressures endemic to high-tech company life. This resiliency could mean graciously handling disappointments (like a good idea rejected as not being cost-effective or being passed over for promotion because management doesn't believe you're sufficiently experienced yet) and demonstrating effectiveness under pressure.

COMPANY LOYALTY

Even if you are convinced that advancement is impossible at your present company, loyalty to the company can be your key to the very advancement you believe impossible. Why? Here's a surprisingly little-known fact: Company loyalty may well be among the top prerequisites for advancement in the high-tech world.

In the present climate of independence and entrepreneurial spirit, job hopping and offshoot companies are as common as silicon chips in Silicon Valley. And not unexpectedly, high-tech management is worried by this high rate of "defection." In a 1983 speech, Intel's chairman Gordon Moore warned that Intel expects "the highest standards of loyalty" from its workers. Going further, Moore added that he considered disloyalty "unethical and wrong."

I quote from Moore not to advise you to plan on a lifetime career at one company, but to illustrate the importance of showing company loyalty *while* you are employed by a particular company. Obviously, what is good policy for a company chairman and for the success of a company as a whole may not always be in an employee's best interests. If you are fired by entrepreneurial spirit or unappreciated at your present company, company loyalty would hold you back from deeper professional and personal satisfactions and greater financial rewards. But if you're not entrepreneurially inclined—or not ready to set off on your own just yet—and eager to advance within your present company, listen to Robert Half's advice.

Half is president and founder of Robert Half, Inc. in New York, a large data processing recruiter. He stresses that getting ahead has a great deal to do with company loyalty. Too often, says Half, people in computer industries "have a love affair with the machinery or the software and have absolutely no loyalty to the company." Those who are concerned with building a career should have a "love affair with the company," advises Half.

This means helping the company make money by being more efficient or developing a money-saving procedure, not leaving to found one's own company or to join another defector's. This, of course, is totally dependent on the likelihood of your being sufficiently rewarded.

At MicroPro International, Inc., a software firm in San Rafael, California, chief programmer Diane Hajicek believes that her talents and company loyalty are well rewarded. She has equity in the company and professional independence. MicroPro recognizes that its success depends on making valued employees like Hajicek happy, and continues to compensate Hajicek well by increasing her stock in the company. Does your company offer stock to employees as part of a compensation program?

Siggy Kauefer, now director of the Paperless Office in Washington, D.C.—a demonstration and educational facility and microprocessor sales outlet—spent three years working as a secretary in one company. She worked her way up to word processing center supervisor, trading on her increasing skills and company loyalty. Her title was tied to her boss's title until she was selected to be WP center supervisor. She broke into management by positioning herself as a loyal employee.

Company loyalty can be compared to patriotism. And not all individuals are capable of demonstrating such devotion. You may be too independent or too much of a maverick (an indication that entrepreneurship may be more your cup of tea or that your present company may be the wrong one for you). However, if you're capable of company loyalty, here's how to advertise your valuable faithfulness. (Take note that simply remaining at a company —one's length of service—may not prove anything. Such service may be considered merely a sign that you lack ambition or risk-taking ability, or have high security needs.)

- Voice respect for your company's product or service. Familiarize yourself with its superiority over the competition's. Show enthusiasm.
- Offer ideas for improving the company's product or service or efficiency in formal memos or at meetings.
- Don't exceed limits of credibility. Temper your enthusiasm with good sense.
- Indicate that you are a team player. Share praise and credit with team members.
- Articulate your understanding of how your individual professional role fits into the larger concerns of the company. Be willing to make sacrifices to further company-wide goals. Recognize the larger picture and show that you do. Know your company's markets and clients and what they value. Understand which departments contribute most to the bottom line.

EXTRA WORK

Tackling tough assignments or taking on extra work traditionally has been a valid route to promotion. It calls attention to such laudatory traits as self-directedness, desire to learn and grow, hard work, enthusiasm, eagerness, and reliability. However, whether such tactics will be identified properly—and positively—as a sign of ambition, or improperly, as a sign of one's acceptance of exploitation, depends on many factors:

- Are you being asked to handle two jobs at once without adequate compensation—in pay and title? (Don't accept one without the other. A salary increase may sweeten the deal, but the title will give you the respect and power you'll need to get the job done and advance further.)
- Is the extra work of limited duration—to meet a foreseeable deadline? Is your beyond-the-call-of-duty effort to be compensated with a bonus, or a pat on the back?
- Is the extra work really "extra"? Or is it a bona fide cyclical or periodic part of your job or simply part of the territory? Is it included in your job description?
- Will turning down an assignment or additional responsibility jeopardize your future growth at the company?
- Will the extra work supply you with knowledge or experience you need to attain your career goals? Is it just "busy work" someone has to do?

Weigh your answers before accepting or rejecting extra work.

HEADHUNTERS

Headhunters play a large role in high-tech advancement. As a woman, you would be especially well served by being informed about how they operate. Why should a woman pay special attention? Because many high-tech companies are actively seeking women candidates, especially in managerial positions, as part of affirmative action programs. (For more on these programs, flip back to Chapter 2.) This can provide you with a distinct advantage, *if* your name is known to headhunters. (Headhunters tend to handle positions typically in higher salary ranges—about $30,000—for qualified, experienced candidates.)

Headhunters—or professional or executive recruiters, or executive search consultants—tend to rely on an informal network of contacts to locate suitable candidates. Rarely do individuals come to them in search of new jobs, as is done with lower level job placement companies. (Recruiters will generally accept résumés, but won't necessarily follow up receipt with an interview unless they have a particular position to fill that requires qualifications like yours.)

Headhunters are given job descriptions and requirements by companies, and then they research suitable candidates. How can you get your name on the list of suitable candidates?

- Be quoted in industry publications or have bylined articles or letters to the editor in them, speak at industry conferences, lead workshops, seminars, or panel discussions.
- If contacted by a recruiter to act as a reference for a co-worker or subordinate, offer your own credentials for consideration for future opportunities.
- Follow up recruiter advertisements in industry publications.
- Be an active member of trade and professional associations.
- Make as many contacts as you can within your industry and industry associations.
- Be listed in who's who directories of your industry.

When dealing with headhunters, you should not pay any fee. And you may well benefit from special coaching on what the potential employer is looking for. After all, a recruiter only earns a fee if she or he successfully fills a position.

Respected headhunters operate with complete confidentiality, never jeopardizing your present position. And smart headhunters keep your résumé filed, updated, and periodically reexamined. Headhunters will want to know your professional background, educational background, technical and management skills, present salary level, desired salary, interest in relocating, references, attitudes toward business travel, professional affiliations, major professional accomplishments, and short- and long-term career goals.

It's important to feel that your headhunter is someone you want to represent you. A headhunter is your front person. Prior to your interview with the company, she or he will try to sell you on the basis of your résumé and whatever other aspects of yourself she or he considers marketable. Look for a headhunter who is well recommended, knowledgeable about your industry, particular positions, and companies, and who has many contacts.

The first time a headhunter contacts you, you may well be pleased with your present position. Be that as it may, it can't hurt you to touch base with a recruiter. Be flattered, and listen carefully to the opportunity available. It may be too good to ignore. Recognize as well that a recruiter will often present more than one candidate to fill a position, so don't equate a recruiter's interest with an endorsement or count the job as yours before it's offered. And don't forget to ask yourself if you're really interested in the position—or whether you think your present company offers you a better opportunity or a more satisfying situation. Are you ready for a change?

In an interview with the recruiter, ask:

- How long has the position been open?
- Is it newly created? If not, what happened to the previous job holder? If created, why?
- What is the nature of the department? Is it new or well-established? Growing or rebuilding?
- What is the company management style? Conservative or risk-taking?
- Is the company well-managed and well-financed? Does it occupy a secure niche in the marketplace?
- What is the company's history?
- Have you had previous experience with this company?
- Am I a serious candidate in your eyes?

- What can I expect to happen next?
- What do you think are my strengths? My weaknesses?

Don't insist on knowing who suggested your name if the reference is not forthcoming. And don't expect to be given the name of the company where the opening exists until just before the interview.

A warning: Don't confuse a career *counselor* with an executive *recruiter*. A counselor may help you prepare a revised résumé, develop a list of prospective employers, write letters, and coach you on interviewing techniques—for a fee. A counselor generally does not provide you with specific leads for employment. If you think you need this service to advance, investigate a prospective counselor carefully. Secure a written contract and read the fine print carefully.

OFFICE POLITICS

Some management experts place more emphasis on political skills in professional advancement than on any other success strategy. And honing one's political skills is largely a product of observation, analysis, and experience. Some political skills are organization-specific. Others are more effective for one sex than the other. The major political skills (sometimes called managerial skills) necessary for high-tech advancement are:

- knowing when to suggest new ideas
- knowing how and to whom to suggest new ideas
- being sensitive to power blocs within an organization—knowing who can effect or block change
- being able to influence, persuade, and motivate others
- knowing where the company's priorities lie
- knowing the company's style—and harmonizing with it
- being able to forecast and pinpoint problems or conflicts and to pose and implement solutions or defuse explosive situations
- knowing how to call positive attention to yourself

- knowing how to make others feel good about you and to win over detractors, promote admirers, and gain respect of competitors
- knowing whom to trust
- knowing how to network and make connections—and capitalize on them
- knowing when to be a team player and when to be independent
- knowing when and where it's appropriate to demonstrate leadership ability
- knowing how to negotiate raises and promotions
- knowing how to trade favors and engage in reciprocity

Many of these skills may sound familiar. You've already met many of them under the guise of separate tools for advancement—like flexibility, likability, and networking. This list can then function as a summary of critical factors in upward mobility.

RISK-TAKING

No chapter on advancing in the high-tech world would be complete without a section on risk-taking. Risk-taking is the name of the game when seeking upward mobility in your career. Yet for many women, risk-taking spells danger, not challenge. Traditionally, women have been raised to steer clear of risks and position themselves on a safe path, where they can make safe choices.

It's true that taking risks will expose you to the possibility of negative consequences. If you ask for a promotion from programmer to systems analyst, you may be turned down. If you take on a new high-tech product as sales rep, you may not be able to meet your quota. On the other hand, asking for a promotion may get you one and taking on a new product might result in an explosion of sales and commissions. Moreover, avoiding risks may not be a safe policy. Your old product may become obsolete;

your programming position may prove a dead end. In high-tech careers, change is too frequent for you to copy the ostrich and hide your head in the sand.

Amy Wohl, president of a successful office automation consultancy, was employed by McGraw-Hill for five years. She was promoted three or four times until she was running all of their office systems programs as an executive editor. The next step was a vice-presidency. Wohl took stock of her position and realized that she was so well-known outside of her company, in the information industry, that she could make much more money as a full-time consultant. She interviewed with many major consulting companies, where she was offered very high salaries—much more than what she was earning in publishing. She realized that she could earn much more by going out on her own, but she hesitated. Why? Wohl attributes her hesitation to fear of failure. It took her time to see that it was OK to fail, that she could accept not succeeding as long as she understood what she'd do if she failed (look for employment as a consultant at a major company). Three-and-a-half years after starting her company, Advanced Office Concepts, Inc., Wohl is happy with her decision and assesses the risk she took as worthwhile.

Risk-taking means embracing opportunities. It can be frightening and anxiety-producing. The outcome will be uncertain, out of your control. How to identify a risk as reasonable? Ask yourself:

- What are the alternatives? Other means of reaching your goal? What would happen if you didn't take the risk?
- Do you have the necessary skills?
- How much are you willing or able to risk?
- Do you have a goal clearly in mind?
- Have you received support or will you receive support for taking on the risk, particularly from those not affected by your action? If not, can you handle the lack of support?
- Are you reasonably confident of your success?

- Can you afford to lose? What's the most at stake?
- Whom are you taking the risk for?
- Is the timing right?
- Does it feel right? (Butterflies in your stomach, insomnia, and mild anxiety are all to be expected initially, but despite your body's natural responses, does it feel right deep inside you?)

These questions aren't easy to answer. Do your best; make your decision. You might want to go back to the start of this chapter and take the quiz, "Are You Management Material?" again. And good luck with your career!

High-Tech Entrepreneurship

Are you entrepreneurial material? Do you really want to run your own business? Would you be happy as an entrepreneur? What motivates you in your career? How much do you really know about what's involved in being a successful entrepreneur? The following quiz will help you answer these questions. Choose the answer that best matches your beliefs.

Test Your Entrepreneurial Stuff

1. My idea of a good time is:
   (a) making decisions
   (b) doing nothing
   (c) spending money

2. It's most important to me to be:
   (a) independent
   (b) secure
   (c) challenged

3. I've always been committed to:
   (a) helping others
   (b) attaining my goals
   (c) making money

4. It's more important to me to be:
   (a) in a happy relationship
   (b) a good parent
   (c) successful in my business

5. The word that characterizes me best is:
   (a) organized
   (b) creative
   (c) driven

6. To me power means:
   (a) running things
   (b) being important
   (c) having a great deal of money

7. For a business to survive, I think the most important factor is:
   (a) product or service competitiveness
   (b) marketing strategy
   (c) maintaining cash flow

8. My feeling about compromise is that it's:
   (a) a necessary evil
   (b) unacceptable
   (c) no big deal

9. The most important trait for a good high-tech employee is:
   (a) trustworthiness
   (b) intelligence
   (c) sense of humor

10. Delayed gratification is:
    (a) a necessary evil
    (b) unacceptable
    (c) no big deal

11. To win a fat contract for my business the most I would be willing to give up would be:
    (a) occasional weekends
    (b) weekends and sex occasionally
    (c) occasionally weekends, sex, and some freedom

12. In high school I was:
    (a) popular
    (b) an outsider
    (c) somewhere in between

13. My mother and/or father are or have been:
    (a) entrepreneurs
    (b) corporate managers
    (c) other

14. The most significant point about a good product or service to remember is that it:
    (a) can always make you rich
    (b) might make it if promoted properly
    (c) needs a good business plan

15. On a busy day at work if my son or daughter became ill I would (Note: view this as a hypothetical case even if you have no children or no spouse):
    (a) take the day off
    (b) talk to my spouse and decide who had more time to spare
    (c) arrange for another responsible person to care for the child

16. I am more afraid of failure than I am of:
    (a) success
    (b) snakes
    (c) getting fat

17. I have known failure:
    (a) intimately
    (b) somewhat
    (c) never

18. My feeling about minor details is:
    (a) let someone else handle them
    (b) I better keep my hand in everything
    (c) I'll deal with them as they become important

19. Good employees:
    (a) can spell the difference between a business's success or failure
    (b) are a dime a dozen
    (c) are born not made

20. My motivation to attain business success is to gain:
    (a) money
    (b) power
    (c) freedom

Score the following points for each answer:

1. a: 5	6. a: 5	11. a: 1	16. a: 1
b: 0	b: 1	b: 5	b: 0
c: 1	c: 0	c: 0	c: 5
2. a: 5	7. a: 0	12. a: 0	17. a: 5
b: 0	b: 1	b: 5	b: 1
c: 1	c: 5	c: 1	c: 0
3. a: 1	8. a: 1	13. a: 5	18. a: 1
b: 5	b: 0	b: 1	b: 0
c: 5	c: 5	c: 0	c: 5
4. a: 1	9. a: 5	14. a: 0	19. a: 5
b: 0	b: 1	b: 1	b: 0
c: 5	c: 0	c: 5	c: 1
5. a: 0	10. a: 5	15. a: 0	20. a: 1
b: 1	b: 0	b: 1	b: 0
c: 5	c: 1	c: 5	c: 5

Key

76 to 100: *Very Strong Entrepreneurial Candidate;* this chapter will tell you how to be successful.

51 to 75: *Strong Entrepreneurial Candidate;* you might consider being a consultant or an entrepreneur.

26 to 50: *Possible Entrepreneurial Candidate;* you need to decide why you want to be an entrepreneur and learn more about entrepreneurial skills.

0 to 25: *Unlikely Entrepreneurial Candidate;* this score doesn't mean you have to give up your dream of being an entrepreneur —if that is your dream. However, you are not as natural an entrepreneur as higher scoring individuals. Nevertheless, you too can probably become entrepreneurial material.

ENTREPRENEURIAL PROFILE

To understand why some answers are scored as more or less characteristic of entrepreneurs, you'll need to learn about existing research concerning entrepreneurs. Psychologists, management consultants, business professors, and writers have tackled this area and come up with some interesting findings. Entrepreneurs, according to two Purdue University professors, tend to be one of three types of people:

- craftspeople (like mechanics and artists) who have a personal skill which they exploit to build a business
- growth-oriented individuals (who want to accomplish financial or other personal goals)
- independent types (who want to avoid working for others)

Arnold Cooper and William Dunkelberg at Purdue studied 890 entrepreneurs. They found that half of these individuals had parents who owned businesses. Two-thirds of the business founders started their first companies when they were between the ages of twenty-five and forty. One-third had full-time partners, and partnership businesses tended to do better financially than single-owner businesses.

Businesses run by craftspeople had the lowest growth rate. Growth-oriented individuals tended to have previous management experiences. And 20 percent of those who became entrepreneurs to avoid working for others had either never been in the labor force or had prior work experience in non-profit organizations.

David McClelland, the renowned Harvard researcher and author, wrote that the drive for personal achievement is more important than the drive for money in entrepreneurs. Entrepreneurs also tend to be:

- energetic
- determined

- decisive
- organized
- good leaders
- self-motivated
- self-assertive
- self-confident
- analytical
- socially-oriented
- willing to forego regular hours

A trait not insignificant to entrepreneurial success is a sense of humor. The ability to laugh and put things in perspective is crucial to seeing a business through hard times.

It's been estimated that 10 to 20 percent of the population is more energetic, often more intense, and generally more hungry for control—prime entrepreneurial candidates. John Welsh is director of the Caruth Institute of Owner-Managed Business, which has studied 5,000 entrepreneurs during a twelve-year period. He notes that entrepreneurial types "need or want to direct their own lives." In addition, immigrants are typically driven to prove themselves in their own businesses. In fact, Edward Roberts of MIT notes, "One could argue that the very act of being an immigrant is an entrepreneurial act. The stakes involved with the failure of the geographic move are [however] much greater than in setting up a business."

Immigrating to a new country also frequently instills immigrants with a drive to prove themselves equal—or superior—to their new compatriots. The new kid on the block is often perceived as an outsider and may be unpopular with her or his peers; an immigrant, who may not speak the language, may feel very isolated. This feeling of being an outsider seems to fuel a desire to "show them" and to prove herself or himself. High-tech entrepreneurs Lore Harp and Helga Johnson were both born in Germany. It was certainly true for both women that success was part of the plan.

Entrepreneurs may seem tireless, but frequently they are not workaholics. They work hard to get things done, not for the sake of work itself. And successful entrepreneurs don't rest on their laurels. Many go on to found company after company, and product after product. Evelyn Berezin, who founded Redactron in the sixties is one example. Today she is a partner in a venture capital firm.

David Silver, author of *The Entrepreneurial Life, How to Go For It and Get It,* surveyed more than 400 entrepreneurs about their motivations to succeed in their own businesses. He discovered that:

- a physical ailment in childhood (like asthma or acne) tended to take them out of the mainstream
- a father who was not around the house much—either because of divorce, death, or a second job—triggered a desire to succeed
- there was pressure at home to do better than one's parents

Although Silver believes there is no significant difference between his portrait of male and female entrepreneurs, Lauren Sellers—a high-tech entrepreneur—suggests one adjustment. ". . . One of the things I find unifying among [successful female entrepreneurs] is they played competitive and team sports," Sellers says. Her company, Micro Courseware Corporation, specializes in educational software for training employees in computer use.

A successful entrepreneur also must have an ability to take risks. Starting a new business is always a risk, no matter how well-financed nor how excellent the product or service. An entrepreneur must be able to balance a desire for security against a need to promote company growth.

For high-tech entrepreneurs, recognizing a good opportunity is more significant than having managerial ability, according to Patricia Braden, author of *Technological Entrepreneurship.* And key factors in successful technological innovation, she points out, may well be the founder's personal drive and desire to succeed, rather than technological expertise.

Frequently, successful high-tech entrepreneurs have had role models, were dissatisfied with their jobs before founding their own companies, and felt that they had acquired sufficient wisdom to handle their own businesses successfully. However, many entrepreneurs were still in school when they started their first businesses. You don't have to be working to be bitten by the entrepreneurial bug.

Many high-tech companies originated at universities or private laboratories or at other high-tech companies. The history of high-tech industries is a story of spin-offs. For example, Hewlett-Packard was formed in 1939, within fifty miles of Stanford. Wang Laboratories was started near Harvard. Digital Equipment Corp. formed out of Lincoln Laboratories. Texas A & M University in Austin has a technology-aid program that uses faculty from its engineering and business schools to help entrepreneurs get started. It also actively seeks start-up capital for its students from alumni. And Stanford University's students have traditionally used the school as a springboard to launching high-tech companies.

Brett Kingstone, for one, founded a Palo Alto fiber optics company which is expected to gross an estimated $5 million in 1984. The entrepreneur's age: a tender twenty-three. Other Stanford students have turned university-acquired computer software knowledge and earthquake safety reports into high-profit consulting services. And Stanford, like many other universities today (including the University of Southern California, the University of California at Los Angeles, Baylor, Harvard, Dartmouth, Pennsylvania, and Georgetown), encourages entrepreneurial activity on the part of its students through courses or centers devoted to the subject. (It's important to note that courses in entrepreneurship don't guarantee success, and most entrepreneurs have never taken a course in the subject.)

Do the characteristics noted previously add up to your self-portrait? Do you recognize yourself at all? You needn't match all these traits exactly, but if you have fundamental difficulties with taking risks, achieving personal goals, making large outlays of time and money, or motivating yourself, you may not have the stuff needed for successful entrepreneurship. Not everyone does, nor should everyone. What's necessary is that you know yourself. Understand what motivates you and what your values are. What are your goals? Your life plan? Does running your own business fit in? Or would you be too immobilized by your fear of financial insecurity and your knowledge of business failure rates?

Keep in mind that personal characteristics alone cannot guarantee you business success. You'll need good contacts for financing and for clients or customers. You'll need to maintain good cash flow, focus on sales as well as research and development of your product or service. You'll need to be able to market and promote your product or service adequately, delegate work, hire, train, and supervise employees. You'll need to understand government and legal regulations concerning small businesses, as well as how to handle growth or financial difficulties.

If a business is not particularly *capital intensive* (requiring large financial infusions at the start), you could begin a business without a large investment. This sort of business is becoming increasingly rare in the high-tech world. For businesses that require large R & D (research and development) budgets—like computer products and peripherals, software, and all manufacturing high-tech companies—sufficient money must be there from the start. Especially if the company does not have a quick turn-around time (the time necessary to recoup expenses and make a profit). Fortunately, venture capital money is more readily available for high-tech companies than ever before. And bank loans to businesswomen are not the rare event they were only ten years ago. In addition, the SBA offers special loans just to women. (However, SBA assistance is not available to publishers.) An extended look at financing can be found later in this chapter.

Some businesses are more labor intensive than capital intensive; this means that start-up requires more time and effort than money on the part of the founder. In the following profiles of high-tech entrepreneurs, Barbara Friedman and Sue Currier fit the category of labor intensive enterprises. You'll have to figure out—from the start—which kind of business yours will be, and where you will find the necessary labor (yours alone? work with a partner? hire helpers?) or necessary capital (bank financing? SBA loan? family loan? savings? venture capital?).

Marketing is another critical aspect of entrepreneurship you'll have to consider before start-up. Many high-tech experts consider marketing the key to a company's success or failure. This involves pinpointing the appropriate market for your product or service (families? businesspeople? children? baby boomers?) and devising the best strategy to reach them (selling the product in computer stores, department stores, other specialty stores, or by mail order). Ellen Lapham's story later in this chapter provides an excellent example of marketing considerations.

If running your own high-tech business sounds like something you want to attempt, read on for an in-depth look at some of the many high-tech entrepreneurial opportunities, and some of the women who've seized them. High-tech industries offer an abundance of money-making opportunities. But remember, each opportunity offers the possibility of success—or failure. Be practical *and* determined. And learn from the experiences of the following entrepreneurs—their mistakes and their wisdom.

ANNE GUTHRIE

Anne Guthrie's computer literacy training center in Washington, D.C.—the TechnoLiteracy Center (TLC)—is her first effort at entrepreneurship. Guthrie's background in education and political science equipped her with teaching skills and an understanding of computers and software, but *not* with business management skills. Nevertheless, spurred by a friend's example and the support of her husband and friends, she opened TLC in July 1983. She used a public relations firm to help publicize the school, which offers computer and software use training classes to adults.

"I just knew there was a market," says Guthrie, "and I had to test the waters." As with most new businesses, the first months were difficult. "A lot of people don't see the relevance of training," she explains. "You have to show people the use."

Guthrie knew that her background as a junior-high-school teacher had trained her in breaking things down into small pieces. When employed as a budget analyst for the U.S. government in the late 1970s, she saw her first computer. To her surprise, she realized that no one else in her office wanted to touch it. For Guthrie, it was love at first sight. And not only did Guthrie find the computer a tremendous aid in statistical work, but her efforts to explain how to use the computer to co-workers earned her praise which added to her self-esteem.

In a consultant position, she again earned praise and recognition for her facility with computers. Deciding she could turn her love of computers into a money-making venture, she began teaching courses on combating technophobia at her local YWCA. A talk at the Washington, D.C. branch meeting of the National Association of Women Business Owners followed. A local TV station covered the talk on the eleven o'clock news. Then after leading another technophobia seminar at George Washington University, Guthrie decided the time was ripe for opening a full-time business.

Guthrie admits to nervousness and uncertainty about her venture. "I have no role models in this except my peers," she notes. "Learning to run a business for money is new to me." Yet Guthrie is hanging in there. She's determined to give her business the old college try. Fears, she recognizes, can keep an entrepreneur from succeeding, as well as keep computer owners and users from fully utilizing their computers.

"I spend a lot of time marketing, talking to people, dealing with their anger and frustration [over problems they experience with computers]. And organizations take a long time to make decisions [about arranging for training of their employees]. It's a new commodity on the market. Personal reference makes the sale," Guthrie observes.

Guthrie's business may not make it, but she's determined to hold on through the initial "hand to mouth" existence for as long as she can. What helps her survive as an entrepreneur? An ability to set objectives, a high energy level, and persistence. If these traits were enough to guarantee entrepreneurial success, Guthrie would have it already.

Entrepreneurial success, however, is also related to factors like competition, timing, financing, advertising, pricing, and business plans. And while many entrepreneurs receive their business training on the job, the lessons can come hard. Guthrie's training center must compete with other similar schools in the Washington, D.C. area. That's where financing, location, pricing, marketing, and advertising will decide who has the competitive edge. More on these crucial factors later. Now let's move on to another high-tech entrepreneurial "case history."

ROBIN OEGERLE

Robin Olin Oegerle followed the awarding of her master's degree in music in 1976 with a job as a computer assistant in a Massachusetts police department. She got the job because she knew how to type. After two years she moved to Arizona, where she taught music for a year. Then her career changed direction again. Her next job was as a research analyst/statistician for an Arizona probation department. She started publishing papers on management information systems, which added to her credentials as a data processing expert. And a year later, in 1978, she decided to switch gears again, capitalizing on her research background and statistical experience. She "sold" herself for a new position as a research associate at a small investment consulting firm in Washington, D.C. She was hired to design a very specialized portfolio analysis and performance measurement software package. As the firm grew (the staff increased from one full-time employee to thirty), Oegerle became the director of research and an administrator. After three years, she was maintaining a data base and supervising and editing her staff's reports. And Oegerle was ready for her next move.

"I wanted to be more challenged. I wanted to do more specialized consulting. And I wanted to work more with people, as well as do something on my own," explains Oegerle. "I decided that financial planning services were really hurting. There are a lot of 'financial planners' out there, but they all work for brokerage firms or insurance companies. And they aren't financial planners. They also provide services only to the very wealthy, so I said if I'm going to be different from everybody else, I'm going to do two things.

"I'm not going to sell anything. And second of all, I'm not going to set a limit on what somebody has to earn in order to qualify for services."

In setting up her business this way right from the start, Oegerle firmly set out to claim a market share she considered hungry for her skills. She surveyed the competition and the competition's product and decided how she would compete. While many businesses do set themselves up in direct competition with others (as Anne Guthrie did), Oegerle tried to meet a need in the marketplace she considered unmet.

Oegerle's background includes a solid grounding in business management. She understood the principles of financing a business, maintaining cash flow, and business planning—and she'd practiced what was only theory to many entrepreneurs. In addition, Oegerle set out to diversify her business. Besides working with individual clients, Oegerle designs financial planning software for banks and credit unions who want to offer financial planning. She also offers this service to other financial planners. And Oegerle has written a book to complement her Fall 1984 financial planning software package for home use.

Oegerle has the head and heart of an entrepreneur. It wasn't the money she was after so much as the challenge. She was offered good jobs, but she didn't want them.

"You have to pay your dues," she says. "I've paid my dues for ten years and decided this was the time to do it. I didn't want to get stuck in the same position."

Oegerle sees the following traits as necessary for a successful entrepreneur: common sense, creativity, administrative ability, and fearlessness. She counts creativity highly, but gives fearlessness the overwhelming edge. "I think fear is what causes most people to become paralyzed [as entrepreneurs]. They start worrying about money."

Oegerle admits that starting your own business is risky but believes, "There's actually less stress in owning your own business than in working for someone else, except you tend to worry about money more often. You do what you have to do about it, and you forget it. Or you will become paralyzed and that's what causes the business to fail."

Oegerle's strategy won't work for everyone. Many entrepreneurs need to worry about money to survive. If they stop, they may find themselves facing bankruptcy, or at the least cash flow problems. (Cash flow is the rate that money comes into, and is paid out of, your business. If the two are not in step, you may have difficulty paying your bills.) But her point about the paralysis induced by fear is well-taken. Balancing a good fiscal policy with a natural fear about low cash reserves is a tricky feat. Many business failures can be related to failure at this task.

Oegerle advises an entrepreneur to pay attention to:

- new marketing ideas
- new product ideas

- staying ahead of the competition
- constantly thinking of the business and how to improve it
- publicity

New business owners should expect their incomes to take a nosedive in the first year of operation. Many entrepreneurs take no money out of the business until the second or third year. Some take out only enough to live on, and plough all other profits back into the business. On the average, it takes three to five years to build up a viable business. (The Small Business Administration reports that 50 percent of all new businesses will fail within five years.) The odds are against success—many businesses fail after the first five years, too—but more people are starting businesses than ever before, especially women. (More on this later.)

Oegerle's final word of advice concerns special problems of women entrepreneurs. "I think more women's businesses fail when they have their offices in their homes—because you just get distracted and tend not to work hard enough. And women have a real attitude problem, a guilt complex about their families vs. their work. I still suffer from that to some extent."

Yet Oegerle also counts her sex as an advantage. "I'm non-threatening," she says. "Dealing with somebody's money and finances is a real sensitive thing." She believes men and women alike come to her because it is easier to tell a woman "they haven't done anything about planning their finances, than to tell a man."

Whether you have an advantage because of your sex, or are disadvantaged as a female entrepreneur, you must know yourself. If you can't work successfully at home, you'll need to have an office outside the home. And if you can't juggle priorities successfully, you'll have to choose what sacrifices you're willing to make and whether you are really cut out to be an entrepreneur. Know your strengths as an entrepreneur. Are you skilled in working with people, but deficient in financial know-how (as Guthrie was)? Or are you strong in both aspects of entrepreneurship, while suffering from psychological roadblocks like fear of success or family guilt?

BARBARA FRIEDMAN

Do you think you see an unfilled niche in the marketplace, a need for a product or service you could provide? Are you bored with what you're doing now? Are you prepared to delay gratifications like salary and free evenings and weekends for a time to build a successful business? If you are, you might take Barbara Friedman as a model.

Friedman, president of Computer Science Press in Rockville, Maryland, was driven by a desire to provide a service and to make a significant contribution. After observing the way her husband's book was published and distributed by a major company, Friedman was convinced there was a better way. She noted that her husband and his computer scientist colleagues all had international reputations, yet all had been dissatisfied with their publishers. When Friedman's husband co-wrote a second book, she decided the time was ripe for her to publish it herself.

Friedman was a housewife at the time. While she had taken tests in the late sixties which showed an aptitude for programming, most of her computer science knowledge came from discussions with her husband and his colleagues. Nevertheless, Friedman—like many entrepreneurs—never considered failure. She had a master's in library science, and recognized a "wide-open business opportunity." She deferred all salary for the first five or six years. (Friedman was lucky not to need to draw money from her business in its first years of operation. Many entrepreneurs can't afford to defer salary for that long.)

The business was incorporated within its first year of operation in 1973, while it was run out of her home. (Friedman's son was born in 1973 as well.) The books were stored in the garage. Other business took place in the kitchen and in other rooms of the house. Friedman did all the copy editing, learned how to contract with printers, and followed suggestions from her husband about where to advertise. In 1975 she began attending conferences, and after 2½ years in business, she hired part-time help. Computer Science Press outgrew the Friedman home along the way. The company is now headquartered in 9,000 square feet of space, but Friedman is contemplating another move to larger quarters. Computer Science Press published eighteen books in 1984 with another forty books on its backlist. Friedman has an agreement with both Radio Shack and Apple to publish how-to books for their computers. In addition, Computer Science Press is the exclusive North American distributor of England's Blackwell Scientific Publications. While the company has twenty employees (Friedman's husband is now editor-in-chief), it does not have any sales representatives. Why? "We do heavy mailings to the universities," answers Friedman. "As a result, we don't need a sales rep." That's a highly unusual statement for a publishing executive to make!

Computer Science Press was listed as number 272 on *Inc.* magazine's 1983 list of the top 500 fastest growing companies in the United States. To what does Friedman attribute her company's success? "A tremendous amount of determination and effort." Friedman's company offers competitive book contracts and what she considers superior book promotion and distribution. Larger publishing companies can't offer the same attention to specialized or technical computer books, Friedman says, because the appeal of these books isn't broad enough to warrant the expenditure for large companies.

ESTHER DYSON

In the high-tech world, publishing often means newsletters, not books. And entrepreneurial activity often wears misleading "clothing," and draws comers from less than direct paths. Esther Dyson, now president and owner of EDventure Holdings, Inc., graduated from Harvard with a B.A. in economics. Her first professional job took her to a news agency in London for six months. Then a return to the U.S. found Dyson in New York, taking a night school course in accounting. The Harvard career placement office produced an interview for a market researcher analyst position, and Dyson's career began to move.

Dyson's professional background includes a three-year stint at *Forbes* magazine, where she progressed from researcher to reporter. A 2½ year tenure at a Wall Street securities firm as an associate research analyst added to her knowledge of corporate finance. The position was created for her, and Dyson took a below-market wage to get her foot in the door. She learned market research at a small company and became marketable in this area quickly. Her specialty was small high-tech companies. (Although both of Dyson's parents are scientists, she didn't have an interest in a technical career herself—although she did have an aptitude.)

Her next career move was to a large brokerage house, where she became a specialist in software. After 2½ years she was bored. (By now, a light should go off in your head when you read the word "bored." It's almost a sure sign of a budding entrepreneur!) Ben Rosen, who was a personal contact, ran a newsletter about the computer industry that he started in the 70s. Dyson came aboard Rosen Research Inc. as editor of the newsletter. In 1983 she purchased the newsletter and became owner and president. She later changed the company's name to EDventure Holdings, Inc. She also runs yearly conferences, consults, and employs three people.

Judging by Dyson's example, high-tech on-the-job experience counts for far more than degrees or educational credentials. "Two years' experience in a good job is probably more useful than an MBA," she says. She believes Americans are too credentials conscious. What counts for an entrepreneur is being able to do the job. "I don't feel hampered by not having an MBA," Dyson adds.

Industry analysts—who may be consultants, regular employees, or entrepreneurs—may have degrees in economics (like Dyson), journalism, electrical engineering, computer science, or business. What's more necessary than any degree, says Dyson, are a knack for marketing, aggressiveness, intelligence, and quickness. As a woman industry analyst, you would be more visible and thus easier to remember. (Dyson believes being a woman is an advantage in sales, but a disadvantage in management.)

Being a member of many trade groups helps Dyson add to her network of contacts for potential clients and even potential employees. And her management experience when working for others helped her learn management techniques. "The final place to learn management techniques is on the job," says Dyson. "You can't learn management totally in school or by watching someone before you."

Dyson counts the desire to get something done as the most significant motivation for entrepreneurial success. An entrepreneur's goal, she says, should be to run a company effectively. "There is usually no need to grab authority," she adds. "My people want me to tell them what to do."

What the computer industry needs are "marketing people who understand the industry," Dyson believes. Managers or owners of high-tech companies must have an appreciation of the importance of research and development as well as know how to handle:

- fast growth
- technical and non-technical staff
- constant flux
- heavy turnover of employees
- lack of good market research from which to make business decisions

Dyson's come a long way from a news agency in London. The best explanation for her career success? She took control of her career. Boredom doesn't create entrepreneurs, it merely triggers entrepreneurial drive. The rest is up to individuals. And hard work may be the single most important component of entrepreneurial success.

ELLEN LAPHAM

As president and founder of Syntauri Corporation in Los Altos, California, Ellen Lapham is more than a committed entrepreneur. She's a true believer. "When I discovered the Syntauri synthesizer I said, 'Wow! This is what I want to do with my life.' It's fun because your heart and your guts are engaged, as well as your mind."

What product earns such rave reviews from its manufacturer? It's an orchestra in a keyboard. Using a split keyboard and specially designed software developed by a team of programmer/musicians, the alphaSyntauri can play up to eight "instruments" at once and record one sound over another. This can create a multitrack system, a boon to professional musicians in the market for high-tech versatility and flexibility. Herbie Hancock may be the best-known fan of the synthesizer. But Syntauri Corporation also markets its products to music students and families. Marketing consultant David Archambault told *Softalk* magazine, "The desire for a musical instrument in the home hasn't changed. It's still there, as strong as ever. Music is the next major area of computer exploitation. We'll see an increasing demand for a computer-based musical instrument as a home-entertainment device."

In 1979 Lapham met Charlie Kellner, who introduced the alphaSyntauri on his own at the National Computer Conference show in New York (as well as at shows in California and in Chicago). She was immediately taken with the potential of the product and convinced Kellner that she and partner (now husband as well) Scott Gibbs had the talent and know-how to make it successful. Lapham and Gibbs bought the rights to the synthesizer and the names of the product and of the company, and established the corporation in California. They paid Kellner royalties for his input. The initial agreement was written on a napkin.

It's not the best way to negotiate a contract, but it worked for Lapham, Gibbs, and Kellner. All claimed satisfaction with the turn of events. The 1982 sales were over $1 million.

Refining and marketing a new product may be one of the most common origins of high-tech companies. Apple computers were created by young Steven Jobs and Steven Wozniak in a California garage. Lore Harp began Vector Graphic on a shoestring budget, marketing and distributing the product developed by Harp's then husband. Lenore Salzbrunn recognized the potential in PEAR software, associated herself with a partner with technical know-how, and created a successful company. The list is growing daily. Yet manufacturers may well face the most difficulties of any high-tech type company.

These companies are capital-intensive. Start-up money for research and development, planning marketing strategy, advertising, and a sales force runs high. New technology, especially, demands continual demonstration and explanation.

"In the professional market," Ilana Wiedhopf, marketing director, told *Softalk,* "they know a lot about music, but they don't feel comfortable with the computer. In the education market, they love the fact that it has a computer. They love *everything* it has. The barrier is budget. In the home market, it's also economics, although as we ride the trend of increasing computer awareness we will do fine."

A new product must also be backed up with a good support network of sales outlets. Syntauri started sales through computer stores before adding music store outlets. Finding the right venue for sales is crucial to a high-tech company's sales. (Some unusual high-tech companies model their sales plans on Tupperware's techniques of party sales; others follow the path of Mary Kay cosmetics or Avon products.) Location and type of sales outlets may be equally important. Marketing consultant Archambault: "The biggest problem we have now is finding dealers sophisticated enough to understand the computers and the music aspects. . . . The computer people are still learning their *own* business.

They aren't sophisticated enough yet to see music as the next major area."

Although the overwhelming percentage of sales outlets for Syntauri are through computer stores, it's the music stores that have the higher sales volume. They also require a great deal of personal attention in the form of cooperative advertising, local seminars, and point-of-purchase brochures. And quality control is a constant concern, as it is with any complex high-tech piece of equipment.

Lapham admits to mistakes made along the way. "I would have gotten more capital in sooner," she told *Softalk,* echoing many a high-tech entrepreneur. "And I should have gotten my board of directors together sooner. We should have been more efficient with our promotional dollars. And I would have loved to have hired a few people sooner, so we would have a broader software product line today." Nevertheless, Lapham observed, "That's the dilemma of a start-up company. You have to focus."

As for marketing mistakes Wiedhopf added, "Anyone marketing a high-tech product to consumers should forget what is inside or how it works. Tell them *why* they should buy it. What it will do for them."

Lapham: "Understand what cash flow really means. Cash flow is the bottom line for any start-up company. I don't have to worry about return on investment yet, but I sure have to worry about whether I can pay my bills. To have controls in place for cash-flow management is very important."

"You can't afford mistakes or fuzzy thinking in a start-up," Lapham continued in her interview with Dennis Briskin,"especially in high-tech types where the technology is moving—or you are moving it—faster than anyone else. Managing in an environment of constant newness means you have to hire people who are good at self-managing.

"The hardest thing in dealing with a start-up is the people part," claims Lapham. "With technology, if you throw enough time and money at a problem you will probably solve it." Lapham speaks with the confidence and optimism of a natural entrepreneur. "But getting the people who can pull that off, providing the internal working relationships that have them feeling satisfied and happy and feeling good about being a team, that's the trick."

That's why Lapham believes hiring the "absolutely best people you can early on in the game" is critical. "You need people who have functional skills and a certain amount of self-confidence." Barbara Stack, vice-president of Rand Capital Corporation—a venture capital firm in Buffalo, New York—agrees. "The entrepreneur can't do everything. It takes a strong person to be able to give up some of those responsibilities."

Strength—and drive—are two attributes Lapham seems to have in abundance. And one more ability which helps explain the phenomenon of Syntauri Corporation: "The synthesizer lets me be a kid with my own product. I think that's great. It's a lot more fun than business-systems software."

JANELLE BEDKE

Janelle Bedke, co-founder of Software Publishing Corp. in Mountain View, California, is in the business of designing software for personal computers. She also derives a great deal of pleasure from her entrepreneurial success. Moreover, she claims that starting a business with partners is the way to go. "I think it's a lot more fun than starting a business alone," she says.

Bedke also emphasizes the time and energy demanded by a new business. "Starting a business and running it are like starting and raising a family. There isn't really room for both." And while many entrepreneurs might disagree with Bedke's assessment, she's not alone in her belief. (But just to present examples of the opposing view, note that Lenore Salzbrunn, Amy Wohl, Anne Guthrie, Barbara Friedman, and two entrepreneurs to be discussed later, Ellen Math and Helga Johnson, all are married with children. It can be done, though it may be difficult at times.)

Bedke advises "thinking through your priorities. Set realistic goals. Know yourself and what you can do.'' She believes the following are necessary traits for entrepreneurial success:

- hard work
- a vision of where you're headed
- financial goals
- ability to know when to say no
- talent for attracting, motivating, and keeping good employees
- good people skills
- technical skills
- knowledge of the market place

''There'll always be more to do than you can do. You have to know what's most important to do first,'' Bedke advises wisely. For the individual who must always finish what she starts before moving on to the next task, entrepreneurial (or consulting) life could be hellish. Anyone who is easily overwhelmed by the amount of work to be done is not cut out to be a happy entrepreneur.

Bedke's list of necessary traits differs from that of many other entrepreneurs by its inclusion of technical skills. She explains, ''A technical background is better for a move into management. It gives you confidence.'' Bedke has never experienced technophobia. Yet she does not wholly attribute this to her educational background in mathematics and computer science. Instead she notes: ''Part of the fear of technology is the fear that 'I haven't kept up.' '' Bedke has kept up.

In ten years at Hewlett-Packard in California, Bedke moved from software designer to a managerial position in research and development of software. She characterizes the experience as one of ''tremendous learning.'' Yet when the time was ripe, in 1980, Bedke left HP to start Software Publishing Corp. with two colleagues from the company. She devoted herself to working full-time for the new company at no salary for the first year. Her partners kept their jobs, but put in time and money. They focused on the personal computer arena as a high-growth area.

One partner wrote their first software package, while Bedke took responsibility for writing the manual (or documentation), meeting with vendors, and announcing the product. Six months after their first product was out, her partners joined her full-time. After three years in business, the company had grown to employ sixty employees. Sales skyrocketed from $1 million in 1981 to $10 million in 1983.

Bedke cautions that start-up companies need generalists more than specialists, or at the least individuals unafraid to get their hands dirty. "In a start-up company the founders end up doing whatever needs to be done." That means if there's no one else to package the product, take orders, or ship it, the president or vice-president of the company will have to do the job. Bedke was originally the general manager for the start-up. Then she became vice-president of marketing. "As a company becomes more established, functions become more organized and can be divided differently," Bedke explains.

As for financing, Software Publishing Corp.—like many other high-tech companies—was started with labor and cash investments from the founders alone. "Then after orders were flooding in for the product, in the first year, we received venture capital to finance growth." For all of the lessons Bedke learned from the start-up and growth of her company, she resists making general pronouncements. "The industry is so young, there aren't so many hard and fast rules." She sees no advantages or disadvantages to her sex, although she notes, "I was an anomaly when I first started. More and more women have taken on managerial roles, but there are still fewer than there should be." Nevertheless, Bedke advises: "Don't think of yourself as different." Her advice has certainly worked for her.

SUE CURRIER

If you're looking for a commonality of background among the high-tech entrepreneurs profiled in this chapter, you might as well throw in the towel. Sue Currier, president of Softsync, Inc. in New York, is a former model. While she is Australian by birth, and as an immigrant may be more driven to succeed than her American-born counterparts, she was happy in her work. In fact, getting into the software business was more something that happened to her than something she actively sought.

In 1981 Currier decided to make a $1,000 investment, incorporate, and distribute programs developed by an Australian friend for Sinclair ZX80s. She and her Australian partner sent the programs out for reviews to computer magazines and started a mail-order business. In this way—without advertising costs, rent, or sales force costs—the company was more labor-intensive than capital-intensive. The brochures they arranged to put in the boxes of Sinclair computers, with the agreement of Sinclair were simple one-page offset printed flyers. Currier printed stationery and invoices and the company was off and running. Within months the small business was demanding all of Currier's free time.

"I never planned on it being a big business," Currier claims. "More a learning experience. Then it began snowballing." And growth, as for many high-tech companies, became problematical. Currier added another ten programs to her list and arranged for licensing of programs in England. Individuals started submitting programs to her for acceptance, and she paid them royalties in return. In May 1982 Timex and Sinclair asked her to put together a line of software for the Timex 1000, which had more memory (and thus was capable of doing more) than their previous models. By June of the same year Currier gave up modeling. The company, which began in her one-bedroom apartment, grew to employ ten. Currier made the decision to go retail, and in October 1982 moved into an office.

Was Currier an unlikely candidate for entrepreneurship? Perhaps. "I had no experience in business," she remarks. "I didn't know anything about doing anything except balancing my checkbook—which I didn't do very well anyway. And I'm learning as

I go along. And you make mistakes as you go along. But I've got to say it's incredibly stimulating. I don't miss modeling at all."

While Currier's enthusiasm bespeaks a natural entrepreneurial drive, her lack of business background is a problem. "We have cash flow problems periodically because people don't pay on time." In addition, the company has grown so quickly that it's taken over her life. (Her husband has joined the company too.) Echoing Janelle Bedke and others, Currier says, "There's always more work than you can do." Yet sometimes Currier sounds like the business is running her, instead of her running the business. Is that the nature of dealing with a growing company? Is it due to a lack of business management skills?

Currier is conservative about both hiring additional employees and seeking venture capital. "I know from talking to a lot of people, that the first thing people do when they start to grow is they hire too many people too quickly."

As for venture capital, Currier is well aware she is being conservative here too. "A lot of companies are running and taking venture capital and then getting a lot of money quickly. And then they're just hiring the experts and getting flashy officers [to help manage the company] . . . I'm trying to do it on what we make and put back in, which is actually a slower development than a lot of people are choosing." Currier is well aware that in taking venture capital, "you lose a lot of control. You have to sell a lot more software to pay for the money you've taken." She admits, "I may not be thinking big enough in some ways." She is looking into venture capital possibilities.

Currier also knows how volatile her business is. She's seen computers become obsolete in a matter of months. She's had to introduce new product lines practically overnight. Attending trade shows to demonstrate her products is a necessary evil. They're necessary, but they're awfully time-consuming. And now, Softsync, Inc., takes out ads in trade magazines to reach buyers instead of consumers. Point-of-purchase flyers at the dealers reach consumers. Sending sample disks (software) to stores is another marketing technique that really pays off, Currier has discovered. Now the company employs sales reps, and hires mostly free-lancers to do packing.

Currier claims her sex is an advantage "without a doubt." Why? "Because it's a male-dominated business, and they love to talk to women." She learns by asking questions and has found both men and women eager to share their knowledge about computers and software.

The intensive experience Currier describes in day-to-day operation of Softsync seems even crazier than the average high-tech business. Is it because there is even more change in the low-cost computer line, or because of a lack of business control? It's a difficult question to answer. Currier admits, "It's an absurd way to live. . . . I keep saying to all my friends 'in six months I'll be a human. I'll talk to you then.' "

Are Currier's sacrifices typical for high-tech entrepreneurs? "When I talk to people who've been in the business for years, they say the first three years are just an absolute grind till everything stabilizes," she says. But do high-tech industries ever really stabilize?

"We're certainly not stable yet," affirms Currier. "It's just as much a roller coaster as it ever was. It's been fun, but at a certain point it does get a bit tiring. But then I seem to thrive on not knowing what's going to happen." And that may well be the prime ingredient demanded by high-tech entrepreneurship. Do you have it?

ELLEN MATH

A background in teaching is one experience that crops up frequently in the lives of high-tech women. Remember Anne Guthrie, Robin Oegerle, and Grace Hopper, among others? Well, add Ellen Math to the list. Ellen Math spent five years as a junior-high-school teacher before leaving to stay home with her children. During the time she spent out of the salaried work force, Math used her organizational talents as a volunteer. She honed her public relations skills during her seven years as a housewife. In 1977 her husband, Irwin Math, said to her: "I'm this great technical guy who can design anything in the world, but when it comes to running a business and paying the bills, I need you." Ellen Math agreed, and a new family business was born.

Family businesses are not uncommon in high-tech industries and elsewhere. Sue Currier's husband works with her. So does Helga Johnson's, Barbara Friedman's, Ellen Lapham's, and Patty Hausman's boyfriend (whom you'll read about). Does this mean women can't run their businesses without men? NO! What it does mean is that sharing a personal life and a professional life with a spouse or close friend can add to the fun, and to the relationship. It can be problematical, but it needn't be. Not when the partners have confidence in each other, as the Maths—among others—do. A clear division of labor helps as well. "I don't tell him how to run his side," notes Ellen Math, "and he doesn't tell me how to do my advertising."

The Maths are conscious, however, that business talk can take over their marriage. To combat this tendency they schedule regular time away in their second home in the Hamptons. And they try to keep their Thursday night dates sacrosanct. (They eat out in a favorite New York restaurant and go to a Broadway show.) After twenty-one years of marriage, and six years as business partners, they seem to have the arrangement perfected. "We're both very happy with our success, and very proud," agrees Ellen Math.

Annual sales of Math Associates' fiber optics (which are used in telecommunications, communications, closed circuit TV) in 1983 were approximately $1 million. The company expanded from their home-base (where their two children helped in the business) to a 3,500 square foot building, and then again to larger quarters. Math Associates employs eighteen.

"We're still a small business," Math notes. And a successful one, which she attributes to the quality and service they provide. The business has been financed largely from personal savings and by deferring salary. This is not uncommon for family businesses. A research grant from a private investment firm in 1982 enabled the Maths to develop their newest product, closed-circuit television using fiber optics.

"What's very important is to be in a field that's growing," advises Ellen Math. As for her personal satisfaction, she notes: "It's much more exciting than opening a dress store or boutique. There are so many fascinating applications of fiber optics. Even English teachers can understand."

HELGA JOHNSON

Of all the male-dominated high-tech industries, biotechnology is the industry in which women are most considered anomalies. For Helga Johnson, the only woman to head an American biotech company, being a woman has often been a disadvantage. In addition, Johnson has only an undergraduate degree in medical technology, from her native Germany. In an industry that is an offshoot of academic labs, lacking a Ph.D. can lead to serious credibility problems. Yet Johnson, president of Tago, Inc. in Burlingame, California, says: "I didn't have any problems as far as dealing with the scientific world or getting started." The problem concerned financing. In the 1970s women had a hard time obtaining bank loans for their businesses. Johnson had to pledge her home. Now, she observes, "The banks are a little more flexible with women these days."

Johnson's desire to explore improving the technology for producing antibodies was her motivation for starting a business with five partners in 1971. Now only one (silent) partner remains. In the beginning all partners kept their jobs and worked on the new venture on the side. The first year was devoted to planning and starting an animal farm. "We weren't really a big operation. We saw this as a very small entity to make a little extra money. We never dreamt it would grow into anything big," explains Johnson. After they received their first contract, "it became a serious thing." In 1972 Johnson was pregnant and the company acquired a small lab to work in. Each partner put in some money (Johnson's initial investment was $350).

In 1977 Tago, Inc. had only three employees. Today there are fifty-two employees. Annual sales in 1972 were $100,000. In 1983, the figure was approximately $1.82 million. Yet in between, the road was rocky. In 1978 the company almost went under. Securing distribution of their product in Germany (where Johnson spent a year with her husband) saved the company. From then on, Johnson knew the company could survive. Her husband, who has a background in marketing, came aboard full-time then to assist in marketing.

The major problem in running a biotech company is financing, according to Johnson. It's costly to have a lab. "You need a lot of R & D money, and we were growing about 60 percent annually. . . . When your business grows, you become more cash poor. And our business is not quite a quick turnover. To cover back orders and cash receivables, you have to have some cash. We received venture capital in May 1982." Johnson retained complete control of the company until June 1983, when she took the company public (sold stock, thereby giving up some decision-making and company privacy) to help finance further growth. She still owns a substantial percentage of Tago, Inc. However, by going public, the company is more "exposed," and dealing with the public is very time-consuming. The bottom line is the substantial cash infusion earned by going public, which makes the move worthwhile.

Johnson attributes Tago, Inc.'s success to her:

- absolute honesty
- listening to what the scientific community had to tell her about their needs
- not being a supersalesperson

Johnson would always blame herself first if she received a complaint about her company's product. She would ask her customers for help and is convinced that this is what earned her the greatest respect.

Johnson is an unusual entrepreneur. A sign of her compassionate treatment of her employees is the way she sent all fifty-two home before Christmas to prepare as they wished for the holidays. That move left her answering the phones herself, but Helga Johnson—like many successful entrepreneurs—does what has to be done. "It is very difficult to be involved in all phases, keep control, and keep the business going," but Johnson succeeds on the scientific and business fronts. While this may make her a "double threat," she's also one of the nicest and most admirable entrepreneurs succeeding in the high-tech world today.

PATRICIA HAUSMAN

Patricia Hausman bought an Osborne computer in June 1982. A free-lance writer, Hausman was eager to use her new computer for word processing and at first saw that function as her only use for the equipment. Instead she became a self-proclaimed computer junkie. "I was fascinated by all the things it could do," remembers Hausman. In short order, Hausman became somewhat of an expert in what the computer could do. "People started calling me up and asking for advice," she says. Since she worked at home and lived in the suburbs in Maryland, Hausman had felt isolated. The conversations with fellow writers and fellow computer users gave her "psychic gratification." She decided to use her expertise and that of friend, engineer Glenn Marcus, to start a cottage "buying service" business in the fall of 1983. Computer Stuff, operated out of Hausman and Marcus's home, sells computers and computer supplies and equipment (such as disks, paper, printer ribbons, acoustic enclosures, etc.). Hausman and Marcus also advise customers on buying the best equipment and provide excellent support.

Hausman's background is as a nutritionist (she has a master's degree in nutrition and spent seven years as a nutritionist and two years editing and writing a nutrition newsletter). In 1981 Hausman decided to combine her nutritional expertise and writing talent. She is the author of *Jack Sprat's Legacy, The At-a-Glance Nutrition Counter,* and *Foods That Fight Cancer.* And in the course of her book research, Hausman has developed a massive nutrient data base, with the help of Glenn Marcus whose professional background includes data base management, MIS (management of information systems), and cost analysis. During the course of Hausman's research and writing, she used a great deal of supplies. She made local contacts for inexpensive computer supplies, which made starting Computer Stuff relatively easy. In addition, Hausman had computer and business know-how. Hausman and Marcus needed no start-up capital because they receive payment from customers before they need to pay for equipment themselves. And since the business is set up in their own home, they didn't need to expend capital on additional rent.

Computer Stuff is a small business, set up as a sole proprietorship; Hausman and Marcus run it as a side venture. Hausman continues to write—and promote—her books. And Marcus continues at his full-time job as an engineer. Running their own business is rewarding, says Hausman. The contact with customers and suppliers helps combat the natural isolation of being a free-lance writer. Hausman smartly combines her computer and nutritional expertise with her writing and business skills, for two satisfying careers. As a free-lance writer and a small business owner, she is doubly an entrepreneur. And for yet another high-tech entrepreneur, new technology signaled the start of an enterprising relationship.

ELISABETH DE SENNEVILLE

French fashion designer Elisabeth de Senneville uses computer graphics to design clothing. In 1980 she discovered computer portraits while on a trip to New York. She was fascinated by the technology. On her return to Paris, where she runs a boutique in Les Halles, she discovered what she calls "scientific graphics." It's a method of using computers to create "mass art." De Senneville isn't interested in making "pure fashion." Instead, her goal is to design clothing that is easy to wear, functional, beautiful, and representative of our time. Using computer-inspired designs is one of her methods. Many of de Senneville's designs use Velcro fasteners or snaps, and her 1983 collection included designs that look quite reminiscent of computer printouts. "I want to mix art and actuality," she says.

De Senneville likes computers, and seems unafraid of new technology. She allows herself not only to be influenced by new technology in her designs, but to use it as well. She is a most unusual fashion designer, as well as an unusual high-tech entrepreneur.

ENTREPRENEURIAL LESSONS

The enthusiasm and excitement with which high-tech entrepreneurs speak of their companies is almost contagious. The thrill they experience from increasing sales, the pride they deservedly feel in their accomplishments ring clearly in interview after interview. The women profiled in this chapter work hard. They work long hours toward goals that are both personal and professional. Janelle Bedke compares running a business to having a family in its demands on one's time and energy. It's an apt analogy.

Entrepreneurial success doesn't come easily. And it doesn't always come to those who deserve it. But it's far from impossible. If you're an ambitious, independent-minded woman, running your own business may be the best career move you could make. The number of self-employed businesswomen has increased 25 percent in the last five years. And women are becoming entrepreneurs at three times the rate of men.

There are 2.8 million businesses in the U.S. today with women as the sole owners. That's more than four times as many as were owned by women five years ago. And if partnerships or corporations where women are partial owners are added, the total rises to 3.7 million—or 25 percent of all small businesses. About half are estimated to be home-based (which appeals because of the flexibility and lack of commuting it offers for personal life and for child care). A study by the American Management Association for the President's Task Force on Women Entrepreneurs noted that women are more likely to plough money back into their businesses, than take quick profits. And this bodes well for their future success.

Capital is of utmost concern to any woman considering founding her own business. The necessary financial investment will depend on:

- the resources needed
- the amount of time estimated before payback (on the average, it's two years before a small business makes a profit)
- the amount of money needed to support the owner before payback

Partnerships, deferring salary, investing savings, bank or SBA loans, venture capital, and going public are financing options discussed, along with their advantages and disadvantages, in the previous case histories.

While time and money vary in business start-ups, entrepreneurs do seem to share some basic personality characteristics. That's what the quiz at the opening of this chapter was designed to test. Now you can see why traits like self-motivation, independence, and flexibility are important. Values like a greater desire for personal achievement than for money augur well for entrepreneurial success. Yet entrepreneurs must be interested in and attentive to money as well if they want to succeed. They must be able to see themselves and their companies in the long term, as well as in the short term. And they must believe fiercely in their ideas and in their company's potential for success.

What sacrifices are you willing to make? While it's less common for today's young female entrepreneurs to give up lasting relationships and children for their businesses, a life filled with marriage, children, and successful entrepreneurship does require a delicate balancing act. Some might consider the task worthy only of a superwoman. For others, it's the only satisfying way to live. Where do you fall along this continuum? Do you have enough energy and drive to succeed at all your endeavors? Can you accept failure should you have to? (Most successful entrepreneurs face failure head-on before they embrace success.)

Are you determined, decisive, and well-organized? Do you consider yourself self-confident and assertive? Can you be aggressive when necessary? Do you have leadership ability? Are you gifted with a talent for analytic and intuitive thinking? Are you blessed with a sense of humor, an understanding of priorities, to help you survive the inevitable difficult times of running your own business? These are all traits considered crucial for would-be entrepreneurs.

Remember that running your own business can be a stressful undertaking. Ask yourself how much stress you are willing to shoulder. Will your business fit in with your personal goals—in terms of relationships, family life, hobbies, other interests? Can the business offer you the kind of security you need? Will it offer you the type of personal and career growth you desire?

While the successful entrepreneur is bound to put in long hours for uncertain rewards at the start of a new enterprise, the future benefits can far outweigh the potential of other types of employment—*for the right kind of person.* In addition to the financial rewards, the flexibility and the satisfactions of being one's own boss and of personal achievement convince many women (and men) that owning their own businesses is the best path to take. And whatever high-tech career path you take, good luck—and much happiness.

APPENDIX I:

Networks, Professional Societies, Trade Associations, and Support Groups*

Accountants Computer Users
Technical Exchange
Suite 115
9100 Purdue Rd.
Indianapolis, IN 46268

Alliance Against Sexual
Coercion
Box 1
Cambridge, MA 02139

American Association of
University Professors
One Dupont Circle, N.W.
Washington, DC 20036

American Association of
University Women
2401 Virginia Ave., N.W.
Washington, DC 20037

American Chemical Society
1155 Sixteenth St., N.W.
Washington, DC 20036

American Electronics
Association
2680 Hanover St.
Palo Alto, CA 94304

American Federation of
Information Processing
Societies, Inc.
1815 N. Lynn St.
Arlington, VA 22209

American Geological Institute
5205 Leesburg Pike
Falls Church, VA 22041

* Address listed is the national headquarters; many associations have local groups as well.

American Institute of
Aeronautics and
Astronautics
1200 Avenue of the Americas
New York, NY 10019

American Institute of
Professional Geologists
Suite 103
7828 Vance Dr.
Arvada, IO 80003

American Medical Women's
Association, Inc.
465 Grand St.
New York, NY 10002

American Medical Writer's
Association
Suite 410
5272 River Rd.
Bethesda, MD 20816

Association for Better
Computer Dealers
861 Corporate Dr.
Lexington, KY 40503

American Society for
Information Science
1010 Sixteenth St., N.W.
Washington, DC 20036

American Society for Medical
Technology
330 Meadowfern Dr.
Houston, TX 77067

American Society for
Microbiology
1931 I St., N.W.
Washington, DC 20036

American Society for Training
and Development
One Dupont Circle, N.W.
Washington, DC 20036

American Society of
Computer Dealers
3500 Southland Center
Dallas, TX 75201

American Society of
Mechanical Engineers
345 E. 47th St.
New York, NY 10017

American Society of
Professional and Executive
Women
1511 Walnut St.
Philadelphia, PA 19102

American Statistical
Association
806 Fifteenth St., N.W.
Washington, DC 20005

American Woman's Economic
Development Corp.
1270 Avenue of the Americas
New York, NY 10020
(800) 442-AWED (in NY);
outside NY (800) 222-
AWED

Association for Computational Linguistics
c/o Dr. D. E. Walker
SRI International
Menlo Park, CA 94025

Association for Computers and the Humanities
Department of English
University of Minnesota
Minneapolis, MN 55455

Association for Computing Machinery
11 W. 42 St.
New York, NY 10036

Association for Educational Data Systems
1201 16th St., N.W.
Washington, DC 20036

Association for Systems Management
24587 Bagley Rd.
Cleveland, OH 44138

Association for Women in Computing
P.O. Box 2293 Grand Central Station
New York, NY 10163

407 Hillmoor Dr.
Silver Spring, MD 20901

c/o Linda Taylor
3572 Greenfield Ave.
Los Angeles, CA 90034

41 Strawberry Circle
Mill Valley, CA 94941

Association for Women in Mathematics
Women's Research Center
828 Washington St.
Wellesley College
Wellesley, MA 02181

Association for Women in Science
Suite 1122
1346 Connecticut Ave., N.W.
Washington, DC 20036

Association of Computer Programmers and Analysts
c/o Cate Corp. Suite 808
1180 Sunrise Valley Dr.
Reston, VA 22091
or
P.O. Box 428
Greenbelt, MD 20770
or
294 Main St.
East Greenwich, RI 02818

Association of Computer Users
P.O. Box 9003
Boulder, CO 80301

Association of Data Communications Users
P.O. Box 20163
Bloomington, MN 55420

Association of Data Processing Service Organizations
Suite 300
1300 N. Seventeenth St.
Arlington, VA 22209

Association of Information Managers
Suite 400
316 Pennsylvania Ave., S.E.
Washington, DC 20003

Association of Information Managers for Savings Institutions
Suite 2221
111 E. Wacker Dr.
Chicago, IL 60601

Association of Public Data Users
Princeton University Computer Center
87 Prospect Ave.
Princeton, NJ 08544

Association of Women in Architecture
7440 University Dr.
St. Louis, MO 63130

Black Data Processing Associations
P.O. Box 7466
Philadelphia, PA 19101

Business and Professional Women's Foundation
2012 Massachusetts Ave., N.W.
Washington, DC 20036

Catalyst
114 E. 60th St.
New York, NY 10022

Center for Women's Policy Studies
Suite 508
2000 P St., N.W.
Washington, DC 20036

Common (IBM business computer users)
Suite 1717
435 N. Michigan Ave.
Chicago, IL 60611

Comparable Worth Project
488 41st St., No. 5
Oakland, CA 94609

Computer-Aided Manufacturing International
Suite 1107
611 Ryan Plaza Dr.
Arlington, TX 76011

Computer Dealers and Lessors Association
1212 Potomac St., N.W.
Washington, DC 20007

Computer and Business Equipment Manufacturers Association
1818 L St., N.W.
Washington, DC 20036

Computer Law Association
c/o Daniel T. Brooks
6106 Lorcan Ct.
Springfield, VA 22152

Computertown USA!
P.O. Box E
Menlo Park, CA 94025

Data Entry Management Association
P.O. Box 3231
Stanford, CT 06905

Data Processing Management Association
505 Busse Highway
Park Ridge, IL 60068

Electronic Data Processing Auditors Association
373 S. Schmale Rd.
Carol Stream, IL 60187

Equal Employment Opportunity Commission
1800 G St., N.W.
Washington, DC 20006

Federally Employed Women
Suite 821
1010 Vermont Ave., N.W.
Washington, DC 20005

Federation of Organizations for Professional Women
Suite 403
2000 P St., N.W.
Washington, DC 20036

GUIDE International (IBM large-scale user group)
111 E. Wacker Dr.
Chicago, IL 60611

Independent Computer Consultants Association
Box 27412
St. Louis, MO 63141

Institute for Certification of Computer Professionals
35 E. Wacker Dr.
Chicago, IL 60601

Institute of Electrical and Electronics Engineers (IEEE)
345 E. 47th St.
New York, NY 10017

IEEE Computer Society
Box 639
Silver Spring, MD 20901

Instrument Society of America
Box 12277
57 Alexander Dr.
Research Triangle Park, NC 27709

International Organization of
Women in
Telecommunications
13450 Maxella Ave #291
Marina Del Ray, CA 90291

Microcomputer Managers
Association
c/o Alan Gross
International Paper Co.
Room 36-3
77 W. 45th St.
New York, NY 10036

National Alliance of Home-
Based Businesswomen
P.O. Box 95
Norwood, NJ 07648

National Association for
Professional Saleswomen
2088 Morley Way
Sacramento, CA 95825

National Association of Bank
Women
111 E. Wacker Dr.
Chicago, IL 60601

National Association of Black
Women Entrepreneurs
P.O. Box 1375
Detroit, MI 48231

National Association of
Female Executives, Inc.
2 Park Ave.
New York, NY 10016

National Association of
Insurance Women
1847 E. 15th St.
Tulsa, OK 74104

National Association of
Science Writers, Inc.
P.O. Box 294
Greenlawn, NY 11740

National Association of
Women Business Owners
Suite 1400
500 N. Michigan Ave.
Chicago, IL 60611

or

2920 M St., N.W.
Washington, DC 20007

National Association of
Women Lawyers
1155 E. 60th St.
Chicago, IL 60637

National Association of
Working Women (9–5)
1224 Huron Rd.
Cleveland, OH 44115

National Computer Graphics
Association
Suite 601
8401 Arlington Blvd.
Fairfax, VA 22031

National Federation of
Business and Professional
Women's Clubs
2012 Massachusetts Ave.,
N.W.
Washington, DC 20036

National Network of Women
in Sales
P.O. Box 95269
Schaumburg, IL 60195

National Organization of
Women
425 13th St., N.W.
Washington, DC 20005

National Society of
Professional Engineers
2029 K St., N.W.
Washington, DC 20006

National Women's Education
Fund
1410 Q St., N.W.
Washington, DC 20009

National Women's
Employment and Education
P.O. Box 959
Suite 622
118 N. Broadway
San Antonio, TX 78205

National Women's Health
Network
224 Seventh St., S.E.
Washington, DC 20003

National Writers Center
13 Astor Pl., 7th Floor
New York, NY 10003

NY Committee for
Occupational Safety and
Health
New York State Consumer
Protection Board
99 Washington Ave.
Albany, NY 12210

Office and Professional
Employees International
Union
Suite 610
265 W. 14th St.
New York, NY 10011

Office Technology
Management Association
Suite 101
9401 W. Beloit Rd.
Milwaukee, WI 53227

Office Technology Research
Group
Box 65
Pasadena, CA 91102

Society for Computer
Applications in Engineering,
Planning and Architecture
358 Hungerford Dr.
Rockville, MD 20850

Society for Industrial and Applied Mathematics
Suite 1405
417 S. 17th St.
Philadelphia, PA 19103

Society for Information Display
654 N. Sepulveda Blvd.
Los Angeles, CA 90049

Society for Management Information Systems
One Illinois Center
Suite 600
111 E. Wacker Dr.
Chicago, IL 60601

Society for Women Engineers
345 E. 47th St.
New York, NY 10017

Society of Computer Simulation
P.O. Box 2228
La Jolla, CA 92038

Society of Office Automation Professionals
233 Mountain Rd.
Ridgefield, CT 06877

Southern California Technology Executives Network (SoCalNET)
12011 San Vacente Blvd.
Brentwood, CA 94513

Women and Mathematics Education
Box 831
Prescott, AZ 86302

Women Entrepreneurs
3061 Fillmore St.
San Francisco, CA 94123

Women in Cable
Suite 703
2033 M St., N.W.
Washington, DC 20036

Women in Cell Biology
Department of Biological Chemistry
School of Medicine
University of California
Davis, CA 95616

Women in Communications, Inc.
P.O. Box 9561
Austin, TX 78766

Women in Data Processing
310 Madison Ave.
New York, NY 10017
or
P.O. Box 8117
San Diego, CA 92102

Women in Information Processing
Lock Box 39173
Washington, DC 20016

Women in Management
525 N. Grant
Westmont, IL 60559

Women in Sales Association
21 Cleveland St.
Valhalla, NY 10595

Women in Science and Engineering
c/o Dr. Miriam Schweber
22 Turning Hill Rd.
Lexington, MA 02171

Women in the Arts Foundation
325 Spring St., Room 200
New York, NY 10013

Women's Bureau
U.S. Department of Labor
14th St. and Constitution Ave., N.W.
Washington, DC 20210

Women's Computer Literacy Project
1195 Valencia St.
San Francisco, CA 94110

Women's Equity Action League
733 Fifteenth St., N.W.
Washington, DC 20005

Women's Occupational Health Resource Center
School of Public Health
60 Haven Ave., B-1
Columbia University
New York, NY 10032

Wonder Woman Foundation
75 Rockefeller Plaza
New York, NY 10019

Working Women's Institute
593 Park Ave.
New York, NY 10021

APPENDIX II

Additional Reading

1 *An Overview*

Academy of Sciences. *Expanding the Role of Women in the Sciences.* New York: Academy of Sciences, 1979.

American Association for the Advancement of Science. *Covert Discrimination and Women in the Sciences.* Boulder, CO: Westview Press, 1978.

American Association for the Advancement of Science. *Women and Minorities in Science.* Boulder, CO: Westview Press, 1982.

Chinn, Phyllis Zweig. *Women in Science and Mathematics.* Arcata, CA: American Association for the Advancement of Science, 1979.

Climbing the Ladder. Washington, DC: National Academy Press, 1983.

Cole, Jonathan R. *Fair Science.* New York: Free Press, 1979.

Davis, Mary Lee. *Women in Science and Medicine.* Minneapolis: T. S. Denison, 1976.

Deken, Joseph. *The Electronic Cottage.* New York: William Morrow, 1982.

Evans, Christopher. *The Making of the Micro: A History of the Computer.* New York: Van Nostrand Reinhold, 1981.

Fins, Alice. *Women in Science.* Skokie, IL: VGM Career Horizons, 1979.

Hanson, Dirk. *The New Alchemists: Silicon Valley and the Microelectronics Revolution.* Boston: Little, Brown, 1982.

Heber, Louis. *Women Pioneers of Science.* New York: Harcourt Brace Jovanovich, 1979.

Hoobler, Icie Gertrude Macy. *Boundless Horizons.* Smithtown, NY: Exposition Press, 1982.

Lantz, Alma E. *Reentry Programs for Female Scientists.* New York: Praeger, 1980.

National Research Council, Committee on the Education and Employment of Women in Science and Engineering. *Climbing the Academic Ladder.* Washington, DC: National Academy of Sciences, 1979.

National Research Council, Committee on the Education and Employment of Women in Science and Engineering. *Women Scientists in Industry and Government.* Washington, DC: National Academy of Sciences, 1980.

Rossiter, Margaret W. *Women Scientists in America.* Baltimore, MD: Johns Hopkins University Press, 1982.

Smith, Walter S. *Science Career Exploration for Women.* Washington, DC: National Science Teachers Association, 1978.

United Nations Institute for Training and Research. *Scientific-Technological Change and the Role of Women in Development.* Boulder, CO: Westview Press, 1982.

Women in Scientific and Engineering Professions. Ann Arbor, MI: University of Michigan Press, 1984.

2 *The Head Start for Women*

Adler, Max K. *Sex Differences in Human Speech.* Hamburg: Buske, 1978.

Conference on Communication, Language, and Sex (First). *Communication, Language, and Sex.* Rowley, MA: Newbury House Publishers, 1979.

Conference on the Sociology of the Languages of American Women. *Proceedings of the Conference on the Sociology of the Languages of American Women.* San Antonio, TX: Trinity University, 1976.

Diagram Group, The. *The Brain: A User's Manual.* New York: Coward, McCann & Geoghegan, 1982.

Eakins, Barbara Westbrook. *Sex Differences in Human Communication.* Boston, MA: Houghton Mifflin, 1978.

Easton, Susan, Joan M. Mills, and Diane Kramer Winokur. *Equal to the Task: How Workingwomen Are Managing in Corporate America.* New York: Seaview Books, 1982.

Farley, Jennie. *Affirmative Action and the Woman.* New York: Amacom, 1979.

Genes and Gender Conference (Second). *Pitfalls in Research on Sex and Gender.* New York: Gordian Press, 1979.

Goldberg, Philip. *The Intuitive Edge.* Los Angeles: J. P. Tarcher, 1983.

Grunwald-Baumrind, Rosalyn M. *The Effects of Sex, Stress, and Personality on Risk-Taking.* 1967.

Hiatt, Mary P. *The Way Women Write*. New York: Teachers College Press, 1977.

Hing, Esther. *Sex Differences in Health and Use of Medical Care*. Hyattsville, MD: U.S. Department of Health and Human Services, Public Health Services, National Center for Health Statistics, 1983.

Hoyenga, Katharine Blick. *The Question of Sex Differences*. Boston, MA: Little, Brown, 1979.

Human Variation. New York: Academic Press, 1978.

Keeves, John P. *Sex Differences in Preparing for Scientific Occupations*. Hawthorn, Australia: Australian Council for Educational Research, 1974.

Key, Mary Ritchie. *Male/Female Language*. Metuchen, NJ: Scarecrow Press, 1975.

Kramarae, Cheris. *Women and Men Speaking*. Rowley, MA: Newbury House Publishers, 1981.

Lakoff, Robin T. *Language and Woman's Place*. New York: Harper & Row, 1975.

Language and Sex. Rowley, MA: Newbury House Publishers, 1975.

Language, Gender, and Society. Rowley, MA: Newbury House Publishers, 1983.

Language, Sex, and Gender. New York: Academy of Sciences, 1979.

McConnell-Ginet, Sally, Ruth Borker, and Nelly Furman, eds. *Women and Language in Literature and Society*. New York: Praeger, 1980.

Psychobiology of Sex Differences and Sex Roles, The. Washington: Hemisphere Publishing Corporation, 1980.

Salk, Jonas. *Anatomy of Reality*. New York: Columbia University Press, 1983.

Second X and Women's Health, The. New York: Gordian Press, 1983.

Seward, Joan Perry. *Sex Differences*. Lexington, MA: Lexington Books, 1980.

Sex Differences in Behavior. Huntington, NY: R.E. Krieger Publishing Company, 1976.

Sex Related Differences in Cognitive Functioning. New York: Academic Press, 1979.

Sorrels, Bobbye. *The Nonsexist Communicator*. Englewood Cliffs, NJ: Prentice-Hall, 1983.

Tavris, Carol and Carole Offir. *The Longest War: Sex Differences in Perspective*. New York: Harcourt Brace Jovanovich, 1977.

Women Look at Biology Looking at Women. Boston, MA: G. K. Hall, 1979.

3 *Technophobia and Training*

ACM. 1982 Administrative Directory. ACM, 1982.

ACM. 1983-84 Graduate Assistantship Directory in the Computer Sciences. ACM, 1983.

ACM Curriculum Committee for Community and Junior College Education. Recommendations and Guidelines for Vocational-Technical Career Programs for Computer Personnel in Operations. ACM, 1981.

ACM Curriculum Committee for Community and Junior College Education; Joyce Currie Little, chairperson. Recommendations for a Two-Year Associate Degree Career Program in Computer Programming. ACM, 1981.

Birney, Robert Charles, Harvey Burdick, and Richard C. Teevan. *Fear of Failure*. New York: Van Nostrand Reinhold, 1969.

Career Opportunity Index. Professional Edition, Western and Southwestern Edition, Vocational Technical Edition. P.O. Box 1878, Huntington Beach, CA 92647.

CDP Exam Guide. New York: Wiley, 1982.

DPMA Model Curriculum for Undergraduate Computer Information Systems Education. Park Ridge, IL: Data Processing Management Association, 1981.

Grundfest, Sandra, editor. *Peterson's Annual Guide to Careers and Employment for Engineers, Computer Scientists, and Physical Scientists*, 1980 edition. Princeton, NJ: Peterson's Guides, 1980.

Hellman, Hal. *Technophobia: Getting Out of the Technology Trap*. New York: M. Evans & Co., 1976.

Morscher, Betsy. *Risk-Taking for Women*. New York: Everest House, 1982.

National Institute of Education, U.S. Department of Education. *Elements of Computer Careers*. Washington, DC: Government Printing Office.

Ritterbush, Philip C., ed. *Technology as Institutionally Related to Human Values*. Washington, DC: Acropolis Books, 1974.

Tobias, Sheila. *Overcoming Math Anxiety*. Boston, MA: Houghton Mifflin, 1980.

U.S. Bureau of Labor Statistics. Occupational Outlook Handbook, 1980–81 edition: Office Machine and Computer Occupations. Washington, DC: Government Printing Office, 1980.

Westlund, John H., ed. *Comp Job*. Chico, CA: Employment Services, 1979.

4 What and Where the Jobs Are

Ayres, Robert U. *Robotics, Applications and Social Implications*. Cambridge, MA: Ballinger Publishing Company, 1983.

Bridwell, Rodger. *High-Tech Investing*. New York: New York Times Books, 1983.

Computer Careers. Skokie, IL: Publications International, 1981.

Cornelius, Hal and William Lewis. *Career Guide for Word Processing*. New York: Monarch Press, 1983.

Data Processing/Information Technology Job Finder, The. Englewood Cliffs, NJ: Prentice-Hall, 1981.

Dobbins, Bill. *"High Tech" Training*. New York: Simon and Schuster, 1982.

Frost & Sullivan. *Computer Graphics Software & Services Market*. New York: Frost & Sullivan, 1979.

Greene, Laura. *Careers in the Computer Industry*. New York: Watts, 1983.

Knight, David C. *Robotics, Past, Present, & Future*. New York: William Morrow, 1983.

Maniotes, John. *Computer Careers; Planning, Prerequisites, Potential*. Rochelle Park, NJ: Hayden Book Company, 1974.

Parkin, Andrew. *Data Processing Management*. Cambridge, MA: Winthrop, 1982.

Rexroad, Robert A. *High Technology Marketing Management*. New York: Wiley, 1983.

Robots Conference (Sixth). *Robots VI*. Dearborn, MI: Robotics International, 1982.

Silverstein, Alvin. *Futurelife, the Biotechnology Revolution*. Englewood Cliffs, NJ: Prentice-Hall, 1982.

Singer, Larry M. *The Data Processing Manager's Survival Manual*. New York: Wiley, 1982.

Tohey, Albert, and Thomas Tilling. *High-tech: How To Find and Profit From Today's New Super Stocks*. New York: Simon and Schuster, 1983.

Vedder, Richard K. *Robotics and the Economy*. Washington, DC: Government Printing Office, 1982.

Winkler, Connie. *The Computer Careers Handbook*. New York: Arco Publishing, 1983.

5 *Getting the Right Job for You*

Atwood, Jerry W. *The Systems Analyst.* Rochelle Park, NJ: Hayden Book Company, 1977.
Brechner, Irv. *Getting into Computers.* New York: Ballantine, 1983.
Briggs, Katherine, and Isabel Briggs Myers. *The Myers-Briggs Type Indicator Manual.* Consulting Psychologists Press, 1975.
Jung, Carl Gustav. *Psychological Types.* Princeton, NJ: Princeton University Press, 1976.
Kurtz, Sheila. *Graphotypes.* New York: Crown Books, 1983.
Makower, Joel. *Office Hazards: How Your Job Can Make You Sick.* Washington, DC: Tilden Press, 1981.
Page, Earle C. *Looking at Type: A Description of the Preferences Reported by the Myers-Briggs Type Indicator.* Gainesville, FL: Center for Applications of Psychological Type, Inc., 1983.
Trembly, Dean. *Learning to Use Your Aptitudes.* San Luis Obispo, CA: Erin Hills Publishers, 1974.

6 *Getting Your Foot in the Door*

Angel, Juvenal Londono. *Effective Résumés for Executives and Specialized Personnel.* New York: Monarch Press, 1980.
Billhartz, Celeste. *The Complete Book of Job Hunting, Finding, Changing.* Akron, OH: Rainbow Collection, 1980.
Catalyst, Inc. *Making the Most of Your First Job.* New York: Putnam, 1981.
Deutsch, Arnold. *The Complete Job Book.* New York: Cornerstone Library, 1980.
Heller, Dorothy K. and June Bower. *Computer Confidence: A Woman's Guide.* Washington, DC: Acropolis Books, 1983.
Holtz, Herman. *Beyond the Résumé.* New York: McGraw-Hill, 1984.
Jackson, Tom. *The Perfect Résumé.* Garden City, NY: Anchor Books, 1981.
Johansen, Ingwald Norman. *Tips on Landing a Federal Government Job.* Annapolis, MD: Job Hunter's Forum, 1976.
Lewis, Adele Beatrice. *How to Write Better Résumés.* Woodbury, NY: Barron's, 1983.
Rosaluk, Warren J. *Throw Away Your Résumé—And Get That Job.* Englewood Cliffs, NJ: Prentice-Hall, 1983.
Shanahan, William F. *Résumés for Computer Professionals.* New York: Arco Publishing, 1983.
———. *Résumés for Engineers.* New York: Arco Publishing, 1983.

———. *Résumés for Technicians*. New York: Arco Publishing, 1983.
Shykind, Maury. *Résumés for Executives and Professionals*. New York: Arco Publishing, 1983.
Summers, Jean. *What Every Woman Needs to Know to Find a Job in Today's Tough Market*. New York: Fawcett Columbine, 1980.
Turbak, Gary. *Action-Getting Résumés for Computer Professionals*. New York: Arco Publishing, 1983.
Weinstein, Bob. *Résumés for Hard Times*. New York: Simon and Schuster, 1982.
Zimmeth, Mary. *The Women's Guide to Reentry Employment*. New York: Charles Scribner's Sons, 1981.

7 *Moving and Advancing*

Back, Ken. *Assertiveness at Work*. London, New York: McGraw-Hill, 1982.
Billings, Karen and David Moursund. *Are You Computer Literate?* Portland, OR: Dilithium Press, 1979.
Bird, Caroline. *Enterprising Women*. New York: Norton, 1976.
Bliss, Edwin C. *Doing It Now*. New York: Charles Scribner's Sons, 1983.
Brown, Linda Keller. *The Woman Manager in the United States*. Washington, DC: Business and Professional Women's Foundation, 1981.
Burka, Jane B. *Procrastination*. Reading, MA: Addison-Wesley Publishing Company, 1983.
Byrne, Joe T. *How to Successfully Ask for a Raise and Promotion*. 1976.
Cannie, Joan Koob. *The Woman's Guide to Management Success*. Englewood Cliffs, NJ: Prentice-Hall, 1978.
Carr-Rufino, Norma. *The Promotable Woman*. New York: Van Nostrand Reinhold, 1983.
Catalyst. *Upward Mobility*. New York: Holt, Rinehart and Winston, 1982.
Collins, Eliza, ed. *Executive Success*. New York: Wiley, 1983.
Collins, Nancy W. *Professional Women and Their Mentors*. Englewood Cliffs, NJ: Prentice-Hall, 1983.
Couger, J. Daniel and Robert A. Zawacki. *Motivating and Managing Computer Personnel*. New York: Wiley, 1980.
Crawford, Jacquelyn S. *Women in Middle Management*. Ridgewood, NJ: Forkher Publishing Corporation, 1977.
Dailey, Charles Alvin. *How to Evaluate People in Business*. New York: McGraw-Hill, 1983.

Davidson, Marylin. *Stress and the Woman Managers*. New York: St. Martin's Press, 1983.

Davis, Maggie S. *Choices of a Growing Woman*. Washington, DC: Acropolis Books, 1983.

Ellis, Albert. *Overcoming Procrastination*. New York: Institute for Rational Living, 1977.

Engel, James F. *Promotional Strategy*. Homewood, IL: R. D. Irwin, 1979.

Farley, Jennie, ed. *The Woman in Management*. Ithaca, NY: ILR Press, New York State School of Industrial and Labor Relations, Cornell University, 1983.

Feingold, S. Norman and Avis J. Nicholson. *Getting Ahead: The Woman's Guide to Career Success*. Washington, DC: Acropolis Books, 1983.

Fenn, Margaret. *Making It in Management*. Englewood Cliffs, NJ: Prentice-Hall, 1978.

Fernandez, John P. *Racism and Sexism in Corporate Life*. Lexington, MA: Lexington Books, 1981.

Fields, Daisy B. *A Woman's Guide to Moving Up in Business and Government*. Englewood Cliffs, NJ: Prentice-Hall, 1983.

Foxworth, Jo. *Boss Lady*. New York: Crowell, 1978.

———. *Wising Up*. New York: Delacorte Press, 1980.

Gardner, Donald G. *The Effects of Sex Stereotypes, Amount of Relevant Information, and Awareness of Organizational Selection Practices on Sex Discrimination for a Managerial Position*. West Lafayette, IN: Institute for Research in the Behavioral, Economic, and Management Sciences, Krannert Graduate School of Management, Purdue University, 1980.

Gordon, Francine E. *Bringing Women into Management*. New York: McGraw-Hill, 1975.

———. *Perspectives on Bringing Women into Management*. Menlo Park, CA: SRI, 1976.

Harragan, Betty Lehan. *Games Mother Never Taught You*. New York: Rawson Associates, 1977.

Hennig, Margaret. *The Managerial Woman*. Garden City, NY: Anchor Press/Doubleday, 1977.

Higginson, M. Valliant. *The Woman's Guide to a Successful Career*. New York: Harper & Row, 1976.

Ilich, John. *Successful Negotiating Skills for Women*. Reading, MA: Addison-Wesley, 1981.

Jennings, Eugene Emerson. *Executive Success: Stresses, Problems, and Adjustment*. New York: Appleton-Century-Crofts, 1967.

Jessup, Claudia. *The Woman's Guide to Starting a Business*. New York: Holt, Rinehart and Winston, 1980.

Jewell, Diana Lewis. *Executive Style*. Piscataway, NJ: New Century Publishers, 1983.

Josefowitz, Natasha. *Paths to Power*. Reading, MA: Addison-Wesley, 1980.

Kellogg, Marion S. *Career Management*. New York: American Management Association, 1972.

Kennedy, Marilyn Moats. *Salary Strategies: Everything You Need to Know to Get the Salary You Want*. New York: Rawson, Wade, 1982.

Knaus, William J. *Do It Now*. Englewood Cliffs, NJ: Prentice-Hall, 1979.

Koch, H. William. *Executive Success: How To Achieve It—How To Hold It*. Englewood Cliffs, NJ: Prentice-Hall, 1973.

Larwood, Laurie. *Women in Management*. Lexington, MA: Lexington Books, 1977.

Lee, Nancy. *Targeting the Top*. Garden City, NJ: Doubleday, 1980.

Lester, Mary. *A Woman's Guide to Starting a Small Business*. New York: Pilot Books, 1981.

Loring, Rosalind. *Breakthrough Women Into Management*. New York: Van Nostrand Reinhold, 1972.

Lynch, Edith M. *The Woman's Guide to Management*. New York: Cornerstone Library, 1978.

MacNeilage, Linda A. *Assertiveness at Work*. Englewood Cliffs, NJ: Prentice-Hall, 1982.

Managerial Career Plateaus. New York: Center for Research in Career Development, Graduate School of Business, Columbia University, 1980.

Oana, Katherine. *Women in Their Own Business*. Skokie, IL: VGM Career Horizons, 1982.

Osborne, Adam. *Running Wild*. Berkeley, CA: Osborne/McGraw-Hill, 1979.

Penn, Margaret. *In the Spotlight*. Englewood Cliffs, NJ: Prentice-Hall, 1980.

Peskin, Dean B. *Womaning, Overcoming Male Dominance of Executive Row*. Port Washington, NY: Ashley Books, 1980.

Pewari, Harish C. *Understanding Personality and Motives of Women Managers*. Ann Arbor, MI: UMI Research Press, 1979.

Place, Irene Magdaline Glazik. *Management Careers for Women*. Louisville, KY: Vocational Guidance Manuals, 1975.

Pletcher, Barbara A. *Saleswoman*. Homewood, IL: Dow Jones-Irwin, 1978.

———. *Travel Sense*. San Francisco: Harbor Publishing, 1980.

Rogalin, Wilma C. *Women's Guide to Management Positions.* New York: Simon and Schuster, 1975.

Rossman, Marge. *When the Headhunter Calls.* Chicago, IL: Contemporary Books, 1981.

Scanlan, Burt K. *Management and Organizational Behavior.* New York: Wiley, 1979.

Schmidt, Warren H. *Managerial Values in Perspective.* New York: AMA Membership Publications Division, American Management Association, 1983.

Scollard, Jeanette Reddish. *No-Nonsense Management Tips for Women.* New York: Simon and Schuster, 1983.

Scruggs, Callie Foster. *Woman in Business.* Mesquite, TX: Ide House, 1981.

Siegelman, Ellen. *Personal Risk.* New York: Harper & Row, 1983.

Stern, Barbara B. *Is Networking for You?* Englewood Cliffs, NJ: Prentice-Hall, 1981.

Stewart, Nathaniel. *The Effective Woman Manager.* New York: Wiley, 1978.

Success Secrets of Successful Women. New York: F. Fell Publishers, 1980.

Thompson, Ann McKay. *Management Strategies for Women.* New York: Simon and Schuster, 1980.

Vetter, Betty M. *Professional Women and Minorities.* Washington, DC: Scientific Manpower Commission, 1975.

Von Oech, Roger. *A Whack on the Side of the Head: How To Unlock Your Mind for Innovation.* New York: Warner Books, 1983.

Watling, Thomas Francis. *Plan for Promotion.* London: Business Books, 1977.

Welch, Mary Scott. *Networking.* New York: Harcourt Brace Jovanovich, 1980.

8 *High-Tech Entrepreneurship*

Baty, Gordon B. *Entrepreneurship for the Eighties.* Reston, VA: Reston Publishing Company, 1981.

Braden, Patricia L. *Technological Entrepreneurship.* Ann Arbor, MI: Division of Research, Graduate School of Business Administration, University of Michigan, 1977.

Brown, Deaver. *The Entrepreneur's Guide.* New York: Macmillan, 1980.

Center for Venture Management. *Symposium on Technical Entrepreneurship.* Milwaukee: Center for Venture Management, 1972.

Christy, Ron. *The Complete Information Bank for Entrepreneurs and*

Small Business Managers. Wichita, KS: Center for Entrepreneurship and Small Business Management, Wichita State University, 1982.

Cohen, William A. *The Entrepreneur and Small Business Problem Solver*. New York: Wiley, 1983.

Entrepreneur's Handbook, The. Dedham, MA: Artech House, 1974.

Entrepreneurship. Lexington, MA: Lexington Books, 1983.

Handy, Jim. *How to Uncover Hidden Business Opportunities That Make Money*. West Nyack, NY: Parker Publishing Company, 1983.

Kaufman, Debra R. *Achievement and Women*. New York: The Free Press, 1982.

Kent, Calvin A. et al., ed. *Encyclopedia of Entrepreneurship*. Englewood Cliffs, NJ: Prentice-Hall, 1982.

Kuriloff, Arthur H. *How to Start Your Own Business . . . and Succeed*. New York: McGraw-Hill, 1978.

McClelland, David Clarence. *The Achieving Society*. New York: Irvington, 1976.

Schollhammer, Hans. *Entrepreneurship and Small Business Management*. New York: Wiley, 1979.

Shilling, Dana. *Be Your Own Boss*. New York: William Morrow, 1983.

Silver, David A. *The Entrepreneurial Life*. New York: Wiley, 1983.

Taffi, Donald J. *The Entrepreneur, a Corporate Strategy for the '80s*. New York: Amacom, 1981.

Van Voorhis, Kenneth R. *Entrepreneurship and Small Business Management*. Boston, MA: Allyn and Bacon, 1980.

Vesper, Karl H. *New Venture Strategies*. Englewood Cliffs, NJ: Prentice-Hall, 1980.

Welsh, John A. *The Entrepreneur's Master Planning Guide*. Englewood Cliffs, NJ: Prentice-Hall, 1983.

Williams, Edward E. *Business Planning for the Entrepreneur*. New York: Van Nostrand Reinhold Company, 1983.

Winston, Sandra. *The Entrepreneurial Woman*. New York: Newsweek Books, 1979.

Young, Dennis R. *If Not for Profit, for What?* Lexington, MA: Lexington Books, 1983.

Index